MW01038510

Number Theory
Through Inquiry

About the cover: The cover design suggests the meaning and proof of the Chinese Remainder Theorem from Chapter 3. Pictured are solid wheels with 5, 7, and 11 teeth rolling inside of grooved wheels. As the small wheels roll around a large wheel with $5 \times 7 \times 11 = 385$ grooves, only part of which is drawn, the highlighted teeth from each small wheel would all arrive at the same groove in the big wheel. The intermediate 35 grooved wheel suggests an inductive proof of this theorem.

Cover image by Henry Segerman
Cover design by Freedom by Design, Inc.

Originally published by
The Mathematical Association of America, 2007.
ISBN: 978-1-4704-6159-1
LCCN: 2007938223

Copyright © 2007, held by the American Mathematical Society
Printed in the United States of America.

Reprinted by the American Mathematical Society, 2020
The American Mathematical Society retains all rights
except those granted to the United States Government.
∞ The paper used in this book is acid-free and falls within the guidelines
established to ensure permanence and durability.
Visit the AMS home page at https://www.ams.org/

10 9 8 7 6 5 4 3 2   25 24 23 22 21 20

AMS/MAA | TEXTBOOKS

VOL **9**

Number Theory Through Inquiry

David C. Marshall and Edward Odell

MAA PRESS

An Imprint of the AMERICAN MATHEMATICAL SOCIETY

Providence, Rhode Island

Council on Publications

James Daniel, *Chair*

MAA Textbooks Editorial Board

Zaven A. Karian, *Editor*

William C. Bauldry
Gerald M Bryce
George Exner
Charles R. Hadlock
Douglas B. Meade
Wayne Roberts
Stanley E. Seltzer
Shahriar Shahriari
Kay B. Somers
Susan G. Staples
Holly S. Zullo

MAA TEXTBOOKS

Contents

0

Introduction

Number Theory and Mathematical Thinking

One of the great steps in the development of a mathematician is becoming an independent thinker. Every mathematician can look back and see a time when mathematics was mostly a matter of learning techniques or formulas. Later, the challenge was to learn some proofs. But at some point, the successful mathematics student becomes a more independent mathematician. Formulating ideas into definitions, examples, theorems, and conjectures becomes part of daily life.

This textbook has two equally significant goals. One goal is to help you develop independent mathematical thinking skills. The second is to help you understand some of the fundamental ideas of number theory.

You will develop skills of formulating and proving theorems. Mathematics is a participatory sport. Just as a person learning to play tennis would expect to play tennis, people seeking to learn to think like a mathematician should expect to do those things that mathematicians do. Also, in analogy to learning a sport, making mistakes and then making adjustments are clear parts of the experience.

Number theory is an excellent subject for learning the ways of mathematical thought. Every college student is familiar with basic properties of numbers, and yet the study of those familiar numbers leads us into waters of extreme depth. Many simple observations about small, whole numbers can be collected, formulated, and proved. Other simple observations about small, whole numbers can be formulated into conjectures of amazing richness. Many simple-sounding questions remain unanswered after literally thousands of years of thought. Other questions have recently been settled after being unsolved for hundreds of years.

Throughout this book, we will continue to emphasize the dual goals of developing mathematical thinking skills and developing an understanding of number theory. The two goals are inextricably entwined throughout and seeking to disentangle the two would be to miss the essential strategy of this two-pronged approach.

The mathematical thinking skills developed here include being able to

- look at examples and formulate definitions and questions or conjectures;
- prove theorems using various strategies;
- determine the correctness of a mathematical argument independently without having to ask an authority.

Clearly these thinking skills are applicable across all mathematical topics and outside mathematics as well.

Note on the approach and organization

Each chapter contains definitions, examples, exercises, questions, and statements of theorems. Definitions are generally preceded by examples and discussion that make that definition a natural consequence of the experience of the examples and the line of thinking presented. We want you to see the development of mathematics as a natural exploration of a realm of thought. Never should mathematics seem to be a mysterious collection of definitions, theorems, and proofs that arise from the void and must be memorized for a test.

Theorem statements arise as crystallized observations. Proofs are clear reasons that the statements are true.

Each chapter concludes with some selective historical remarks on the chapter's content. This is meant to place the ideas on an historical timeline. It is fascinating to see threads begin in antiquity and continue into the 21st century with no clear end in sight.

Chapters one through four present concepts that are used in all the future chapters. Chapter five on cryptography does not contain material that is required for the future chapters. Chapters six, seven, and eight are sequentially dependent. Chapters nine and ten are independent and can be read any time after chapter four. In a semester course, the authors generally treat chapters one through five, using the further chapters for future work and independent study projects.

Number theory contains within it some of the most fascinating insights in mathematics. We hope you will enjoy your exploration of this intriguing domain.

Methods of thought

Methods of thought, proof, and analysis are not facts to be learned once and set aside. They become useful tools as they appear recurrently in different contexts and as you begin to incorporate them into your habits of approaching the unknown.

While looking at numbers and finding patterns among them, it will be natural to develop an understanding of various ways to give convincing arguments. These different styles of proofs will become familiar and logically sound. We do not present these methods of proof in the abstract, but instead you will develop them as naturally occurring methods of stating logically correct reasons for the truth of statements.

Some methods of thought, proof, and analysis are:

- Finding patterns and formulating conjectures.

- Making precise definitions.

- Making precise statements.

- Using basic logic.

- Forming negations, contrapositives, and converses of statements.

- Understanding examples.

- Relating examples to the general case.

- Generalizing from examples.

- Measuring complexity.

- Looking for elementary building blocks.

- Following consequences of assumptions.

- Methods of proof:
    - induction,
    - contradiction,
    - reducing complexity,
    - taking reasoning that works in a special case and making it general.

By the end of the course these abilities and techniques will be natural strategies for you to use in your mathematical investigations and beyond.

We hope you enjoy your inquiry into number theory.

Acknowledgments

We thank the Educational Advancement Foundation and Harry Lucas, Jr. for their generous support of the Inquiry Based Learning Project, which has inspired us and many other faculty members and students. Many of the instructors who tested these materials received mentoring and incentives from the EAF, and we have received support in the writing of this book and other Inquiry Based Learning material. The EAF fosters methods of teaching that promote independent thinking and student creativity, and we hope that this book will make those methods broadly available to many students. We thank the National Science Foundation for its support of this project under NSF-DUE-CCLI Phase I grant 0536839, and Louis Beecherl for his generous support of this work.

Special thanks are also due to the many students and instructors who used earlier versions of this book and who made many useful suggestions. In particular we wish the thank the following faculty members who used drafts of this book while teaching number theory at The University of Texas at Austin: Gergely Harcos, Alfred Renyi Institute of Mathematics; Ben Klaff, The University of Texas at Austin; Deepak Khosla, The University of Texas at Austin; Susan Hammond Marshall, Monmouth University; Genevieve Walsh, Tufts University. We also thank Stephanie Nichols who is a graduate student in mathematics education at The University of Texas at Austin. She took the class, served as a graduate student assistant for several semesters, and is conducting research about the efficacy of this method of introducing students to the ideas of mathematical proof. Thanks also to Professor Jennifer Smith and her students who are doing research in mathematics education that involves inquiry based instruction in the acquisition of mathematical thinking skills.

David Marshall: I thank foremost my coauthors Mike Starbird and Ted Odell for introducing me to the Modified Moore Method style of inquiry based teaching and for mentoring me during my short stay at The University of Texas. The experience was fantastic and has had a profound impact on the way I conduct my classes today. I thank Mike and Ted as well for inviting me to take part in this project. It has been a very enjoyable, educational, and rewarding experience. I thank my wonderful family; my wife, Susan, who has had to listen to me pontificate on all matters number theory for well over a year, and my daughter, Gillian, who always makes coming home the high point of my day.

Edward Odell: Five years ago I spent numerous hours attending Mike Starbird's inquiry based number theory class and then attempting to duplicate his wizardry in my own class. I am forever grateful to Mike for inviting me into this project and for his constant support. Thanks are also due to David, a joy to work with and without whose efforts and guidance this book would still be far from completion. Last but not least I thank my wife Gail for her love and support and my children Holly and Amy for understanding when their dad was busy.

Michael Starbird: Thanks to Ted and David for making the writing of this book an especially enjoyable experience. Their unfailing cheerfulness and good sense made this project a true joy to work on. Thanks also to my wife Roberta, and children, Talley and Bryn, for their constant encouragement and support.

1

Divide and Conquer

Divisibility in the Natural Numbers

How can one natural number be expressed as the product of smaller natural numbers? This innocent sounding question leads to a vast field of interconnections among the natural numbers that mathematicians have been exploring for literally thousands of years. The adventure begins by recalling the arithmetic from our youth and looking at it afresh.

In this chapter we start our investigation of the natural numbers by defining divisibility and then presenting the ideas of the Division Algorithm, greatest common divisors, and the Euclidean Algorithm. These ideas in turn allow us to find integer solutions to linear equations.

The natural numbers are naturally ordered in one long ascending list; however, many experiences in everyday life are cyclical—hours in the day, days in a week, motions of the planets. This concept of cyclicity gives rise to the idea of modular arithmetic, which formalizes the intuitive idea of numbers on a cycle. In this chapter, we will introduce the basic idea of modular arithmetic but will develop the ideas further in future chapters.

As you explore questions of divisibility of integers and questions about modular arithmetic, you will develop skills in proving theorems, including proving theorems by induction.

Definitions and examples

Many people view the natural numbers as the most basic of all mathematical ideas. A 19th century mathematician, Leopold Kronecker, famously said roughly, "God gave us the natural numbers—all else is made by humankind." The natural numbers are the counting numbers to which we were introduced in our childhoods.

Definition. The *natural numbers* are the numbers $\{1, 2, 3, 4, \ldots\}$.

The ideas of 0 and negative numbers are abstractions of the natural numbers. Those ideas extend the natural numbers to the integers.

Definition. The *integers* are $\{\ldots, -3, -2, -1, 0, 1, 2, 3, \ldots\}$.

The basic relationships between integers that we will explore in this chapter are based on the divisibility of one integer by another.

Definition. Suppose a and d are integers. Then d *divides* a, denoted $d\,|\,a$, if and only if there is an integer k such that $a = kd$.

Notice that this definition gives us a practical conclusion from the assertion that the integer d divides the integer a, namely, the existence of a third integer k with its multiplicative property, namely, that $a = kd$. Mathematical definitions encapsulate intuitive ideas, but then pin them down. Having this formal definition of divisibility will allow you to say clearly why some theorems about divisibility are true. Remembering the formal definition of divisibility will be useful throughout the course.

We next turn to a more complicated definition that we will see captures the idea of numbers arranged in a cyclical pattern. For example, if you wrote the natural numbers around a clock, you would put 13 in the same place as 1 and 14 in the same place as 2, etc. That idea is what is formalized in the following definition of congruence.

Definition. Suppose that a, b, and n are integers, with $n > 0$. We say that a and b are *congruent modulo n* if and only if $n\,|\,(a - b)$. We denote this relationship as
$$a \equiv b \pmod{n}$$
and read these symbols as "a is congruent to b modulo n."

We will soon begin with the first set of questions. They come in several different flavors which we roughly categorize as "Theorem" (or "Lemma" or "Corollary"), "Question", or "Exercise." A Theorem denotes a mathematical statement to be proved by you. For example:

Example Theorem. *Let n be an integer. If $6|n$, then $3|n$.*

Then you would supply the proof. For example, your proof might look like this:

Example Proof. Our hypothesis that $6|n$ means, by definition, that there exists an integer k such that $n = 6k$. The conclusion we want to make is that 3 also divides n. By definition, that means we want to show that there exists an integer k' such that $n = 3k'$. Since $n = 6k = 3(2k)$, we can take $k' = 2k$, satisfying the definition for n to be divisible by 3. □

Here's an example that uses a congruence.

Example Theorem. *Let k be an integer. If $k \equiv 7$* (mod 2)*, then $k \equiv 3$* (mod 2)*.*

Example Proof. Our hypothesis that $k \equiv 7$ (mod 2) means, by definition, that $2|(k - 7)$, which, also by definition, means there exists an integer j such that $k - 7 = 2j$. Adding 4 to both sides of the last equation yields $k - 3 = 2j + 4 = 2(j + 2)$. Since $j + 2$ is also an integer, this means $2|(k - 3)$, or $k \equiv 3$ (mod 2), and so the theorem is proved. □

In giving proofs, rely on the definitions of terms and symbols, and feel free to use results that you have previously proved. Avoid using statements that you "know", but which we have not yet proved.

A "Question" is often open-ended, leaving the reader to speculate on some idea. These should be given considerable thought. An "Exercise" is often computational in nature, illustrating the results of previous (or upcoming) theorems. These help you to make sure your grasp of the material is firm and grounded in the reality of actual numbers.

Divisibility and congruence

The next theorems explore the relationship between divisibility and the arithmetic operations of addition, subtraction, multiplication, and division. Frequently a good strategy for generating valuable questions in mathematics is to take one concept and see how it relates to other concepts.

1.1 Theorem. *Let a, b, and c be integers. If $a|b$ and $a|c$, then $a|(b + c)$.*

1.2 Theorem. *Let a, b, and c be integers. If $a|b$ and $a|c$, then $a|(b - c)$.*

1.3 Theorem. *Let a, b, and c be integers. If $a|b$ and $a|c$, then $a|bc$.*

Any theorem has a hypothesis and a conclusion. That structure of theorems automatically suggests questions, namely, can the theorem be strengthened? If we are able to deduce the same result with fewer or weaker hypotheses, then we will have constructed a stronger theorem. Similarly, if we are able to deduce a stronger conclusion from the same hypotheses, then we will have constructed a stronger theorem. So attempting to weaken the hypothesis and still get the same conclusion, or keep the same hypotheses but deduce a stronger conclusion, are two fruitful investigations to follow when we seek new truths. So let's try this strategy with the previous theorem.

When you are considering whether a particular hypothesis implies a particular conclusion, you are considering a conjecture. Three outcomes are possible. You might be able to prove it, in which case the conjecture is changed into a theorem. You might be able to find a specific example (called a counterexample) where the hypotheses are true, but the conclusion is false. That counterexample would then show that the conjecture is false. Frequently, you cannot settle the conjecture either way. In that case, you might try changing the conjecture by strengthening the hypothesis, weakening the conclusion, or otherwise considering a related conjecture.

1.4 Question. *Can you weaken the hypothesis of the previous theorem and still prove the conclusion? Can you keep the same hypothesis, but replace the conclusion by the stronger conclusion that $a^2|bc$ and still prove the theorem?*

If you consider a conjecture and discover it is false, that is not the end of the road. Instead, you then have the challenge of trying to find somewhat different hypotheses and conclusions that might be true. All these strategies of exploration lead to new mathematics.

1.5 Question. *Can you formulate your own conjecture along the lines of the above theorems and then prove it to make it your theorem?*

Here is one possible such theorem. Maybe it is the one you thought of or maybe you made a different conjecture.

1.6 Theorem. *Let a, b, and c be integers. If $a|b$, then $a|bc$.*

Let's now turn to modular arithmetic. To begin let's look at a few specific examples with numbers to gain some experience with congruences modulo a number. Doing specific examples with actual numbers is often a good strategy for developing some intuition about a subject.

1.7 Exercise. *Answer each of the following questions, and prove that your answer is correct.*

   1. Is $45 \equiv 9$ (mod 4)?

   2. Is $37 \equiv 2$ (mod 5)?

   3. Is $37 \equiv 3$ (mod 5)?

   4. Is $37 \equiv -3$ (mod 5)?

You might construct some exercises like the preceding one for yourself until you are completely clear about how to determine whether or not a congruence is correct.

When we gain some experience with a concept, we soon begin to see patterns. The next exercise asks you to find a pattern that helps to clarify what groups of integers are equivalent to one another under the concept of congruence modulo n.

1.8 Exercise. *For each of the following congruences, characterize all the integers m that satisfy that congruence.*

   1. $m \equiv 0$ (mod 3).

   2. $m \equiv 1$ (mod 3).

   3. $m \equiv 2$ (mod 3).

   4. $m \equiv 3$ (mod 3).

   5. $m \equiv 4$ (mod 3).

To understand the definition of congruence, one strategy is to consider the extent to which congruence behaves in the same way that equality does. For example, we know that any number is equal to itself. So we can ask, "Is every number congruent to itself?" The reason that this is even a question is that congruence has a specific definition, so we need to know whether that specific definition allows us to deduce that any number is congruent to itself.

1.9 Theorem. *Let a and n be integers with* $n > 0$. *Then* $a \equiv a$ (mod n).

We will explore several cases where properties of ordinary equality suggest questions about whether congruence works the same way. For example, in equality, the order of the left-hand side versus the right-hand side of an equals sign does not matter. Is the same true for congruence?

1.10 Theorem. *Let a, b, and n be integers with n > 0. If a ≡ b* (mod *n*), *then b ≡ a* (mod *n*).

Again, if *a* is equal to *b* and *b* is equal to *c*, we know that *a* is equal to *c*. But does the definition of congruence allow us to conclude the same about a string of congruences? It does.

1.11 Theorem. *Let a, b, c, and n be integers with n > 0. If a ≡ b* (mod *n*) *and b ≡ c* (mod *n*), *then a ≡ c* (mod *n*).

Note: If you are familiar with *equivalence relations*, you may note that the previous three theorems establish that congruence modulo *n* defines an equivalence relation on the set of integers. In the exercise before those theorems, you described the equivalence classes modulo 3.

The following theorems explore the extent to which congruences behave the same as ordinary equality with respect to the arithmetic operations. We systematically go through the operations of addition, subtraction, and multiplication. Division, as we will see, requires more thought.

1.12 Theorem. *Let a, b, c, d, and n be integers with n > 0. If a ≡ b* (mod *n*) *and c ≡ d* (mod *n*), *then a + c ≡ b + d* (mod *n*).

1.13 Theorem. *Let a, b, c, d, and n be integers with n > 0. If a ≡ b* (mod *n*) *and c ≡ d* (mod *n*), *then a − c ≡ b − d* (mod *n*).

1.14 Theorem. *Let a, b, c, d, and n be integers with n > 0. If a ≡ b* (mod *n*) *and c ≡ d* (mod *n*), *then ac ≡ bd* (mod *n*).

Congruences also work well when taking exponents, as we will see in Theorem 1.18. One way to approach its proof is to start with simple cases and see how the general case follows from them. The following exercises, which are actually little theorems, present a strategy of reasoning known as *proof by mathematical induction*. In the appendix we explore this technique in more detail.

1.15 Exercise. *Let a, b, and n be integers with n > 0. Show that if a ≡ b* (mod *n*), *then $a^2 ≡ b^2$* (mod *n*).

1.16 Exercise. *Let a, b, and n be integers with n > 0. Show that if a ≡ b* (mod *n*), *then $a^3 ≡ b^3$* (mod *n*).

1.17 Exercise. *Let a, b, k, and n be integers with n > 0 and k > 1. Show* *that if $a \equiv b \pmod{n}$ and $a^{k-1} \equiv b^{k-1} \pmod{n}$, then*

$$a^k \equiv b^k \pmod{n}.$$

1.18 Theorem. *Let a, b, k, and n be integers with $n > 0$ and $k > 0$. If $a \equiv b \pmod{n}$, then*

$$a^k \equiv b^k \pmod{n}.$$

At this point you have proved several theorems that establish that congruences behave similarly to ordinary equality with respect to addition, subtraction, multiplication, and taking exponents. To make all these theorems more meaningful, it is helpful to see what they mean with actual numbers. Doing examples is a good way to develop intuition.

1.19 Exercise. *Illustrate each of Theorems 1.12–1.18 with an example using actual numbers.*

You will have noticed that at this point, we have not yet considered the arithmetic operation of division. We ask you to consider the natural conjecture here.

1.20 Question. *Let a, b, c, and n be integers for which $ac \equiv bc \pmod{n}$. Can we conclude that $a \equiv b \pmod{n}$? If you answer "yes", try to give a proof. If you answer "no", try to give a counterexample.*

We will continue the discussion of division at a later point. In the meantime, we find that the concept of congruence and the theorems about addition, subtraction, multiplication, and taking exponents allow us to prove some interesting facts about ordinary numbers. You may already have been told how to tell when a number is divisible by 3 or by 9. Namely, you simply add up the digits of the number and ask whether the sum of the digits is divisible by 3 or 9. For example, 1131 is divisible by 3 because 3 divides $1 + 1 + 3 + 1$. In the next theorems you will prove that these techniques of checking divisibility work.

1.21 Theorem. *Let a natural number n be expressed in base 10 as*

$$n = a_k a_{k-1} \ldots a_1 a_0.$$

(Note that what we mean by this notation is that each a_i is a digit of a regular base 10 number, not that the a_i's are being multiplied together.) If $m = a_k + a_{k-1} + \cdots + a_1 + a_0$, then $n \equiv m \pmod{3}$.

Theorem. *A natural number that is expressed in base* 10 *is divisible by* 3 *if and only if the sum of its digits is divisible by* 3.

Note: An "if and only if" theorem statement is really two separate theorems that need two separate proofs. A good practice is to write down each statement separately so that the hypothesis and the conclusion are clear in each case. We have done that for you in the following case to illustrate the practice.

1.22 Theorem. *If a natural number is divisible by* 3, *then, when expressed in base* 10, *the sum of its digits is divisible by* 3.

1.23 Theorem. *If the sum of the digits of a natural number expressed in base* 10 *is divisible by* 3, *then the number is divisible by* 3 *as well.*

When we have proved a theorem, it is a good idea to ask whether there are other, related theorems that might be provable with the same technique. We encourage you to find several such divisibility criteria in the next exercise.

1.24 Exercise. *Devise and prove other divisibility criteria similar to the preceding one.*

The Division Algorithm

We next turn our attention to a theorem called the Division Algorithm. Before we state it, we point out a fact about the natural numbers that is obviously true. In fact, it's so obvious that it is an axiom, meaning a statement that we accept as true without proof. The reason that we can't really give a proof of it is that we have not really defined the natural numbers, but are just using them as familiar objects that we have known all our lives. If we were taking a very abstract and formal approach to number theory where we defined the natural numbers in terms of set theory, for example, the following statement might be one of the axioms we would use to define the natural numbers. Instead, we will just assume that the following Well-Ordering Axiom for the Natural Numbers is true.

Axiom (The Well-Ordering Axiom for the Natural Numbers). *Let S be any non-empty set of natural numbers. Then S has a smallest element.*

Since we are accepting this fact as true, you should feel free to use it whenever you wish. The value of this axiom is that it sometimes allows us to pin down the reason why some assertion is true in a proof. Here is an example of how you might use the Well-Ordering Axiom for the Natural Numbers.

Example Theorem. *For every natural number n there is a natural number k such that 7k differs from n by less than 7.*

Example Proof. We could let S be the set of all numbers $7i$, where i is a natural number, such that $7i$ is greater than or equal to n. By the Well-Ordering Axiom for the Natural Numbers, S has a smallest element, call it $7j$. Then $7j$ differs from n by less than 7, otherwise $7(j-1)$ would be a smaller element of S. □

This example gives the flavor of how the Well-Ordering Axiom for the Natural Numbers is used; namely, we define an appropriate non-empty set of natural numbers and then look at that set's smallest element to deduce something we want. You might consider using the Well-Ordering Axiom for the Natural Numbers in proving the Division Algorithm below.

The Division Algorithm is a useful observation about natural numbers. Surprisingly often it captures exactly what we need to know to prove theorems about integers. After reading it carefully, you will see that it captures a basic property about ordinary division.

Theorem (The Division Algorithm). *Let n and m be natural numbers. Then (existence part) there exist integers q (for quotient) and r (for remainder) such that*

$$m = nq + r$$

and $0 \le r \le n - 1$. Moreover (uniqueness part), if q, q' and r, r' are any integers that satisfy

$$m = nq + r$$
$$= nq' + r'$$

with $0 \le r, r' \le n - 1$, then $q = q'$ and $r = r'$.

As usual, it is useful to look at some examples with actual numbers to understand the statement.

1.25 Exercise. *Illustrate the Division Algorithm for:*

 1. $m = 25, n = 7$.

2. $m = 277$, $n = 4$.

3. $m = 33$, $n = 11$.

4. $m = 33$, $n = 45$.

1.26 Theorem. *Prove the existence part of the Division Algorithm.*
(Hint: Given n and m, how will you define q? Once you choose this q, then how is r chosen? Then show that $0 \le r \le n-1$.)

1.27 Theorem. *Prove the uniqueness part of the Division Algorithm.*
(Hint: If $nq + r = nq' + r'$, then $nq - nq' = r' - r$. Use what you know about r and r' as part of your argument that $q = q'$.)

The following theorem connects the ideas of congruence modulo n with remainders such as occur in the Division Algorithm. It says that if the remainders are the same when divided by the modulus, then the numbers are congruent.

1.28 Theorem. *Let a, b, and n be integers with $n > 0$. Then $a \equiv b$ (mod n) if and only if a and b have the same remainder when divided by n. Equivalently, $a \equiv b$ (mod n) if and only if when $a = nq_1 + r_1$ ($0 \le r_1 \le n-1$) and $b = nq_2 + r_2$ ($0 \le r_2 \le n-1$), then $r_1 = r_2$.*

Greatest common divisors and linear Diophantine equations

The divisors of an integer tell us something about its structure. One of the strategies of mathematics is to investigate commonalities. In the case of divisors, we now move from looking at the divisors of a single number to looking at common divisors of a pair of numbers. This strategy helps to illuminate relationships and common features of numbers.

Definition. A *common divisor* of integers a and b is an integer d such that $d \mid a$ and $d \mid b$.

Once we have isolated a definition such as common divisor, we proceed to explore its implications. The first question involves how many common divisors there are to a pair of integers.

1.29 Question. *Do every two integers have at least one common divisor?*

1.30 Question. *Can two integers have infinitely many common divisors?*

The greatest common divisor is a concept that plays a central role in the study of many of our future topics.

Definition. The *greatest common divisor* of two integers a and b, not both 0, is the largest integer d such that $d\,|\,a$ and $d\,|\,b$. The greatest common divisor of two integers a and b is denoted $\gcd(a,b)$ or more briefly as just (a,b).

One indication of the centrality of the concept of greatest common divisor is that it has two different notations including the extremely simple notation (a,b). You might think that this notation would be confusing because it is the same notation as for an interval on the real line; however, in the context of number theory, (a,b) always stands for the greatest common divisor.

Having more divisors in common shows some commonality between numbers, but having almost no common divisors indicates that the numbers do not share many factors. A pair of numbers that have no non-trivial common divisors have a special role to play and consequently are given a name, relatively prime.

Definition. If $\gcd(a,b) = 1$, then a and b are said to be *relatively prime.*

As usual, a good way to develop intuition about a concept is to investigate some specific examples.

1.31 Exercise. *Find the following greatest common divisors. Which pairs are relatively prime?*

1. $(36, 22)$

2. $(45, -15)$

3. $(-296, -88)$

4. $(0, 256)$

5. $(15, 28)$

6. $(1, -2436)$

The next theorems explore conditions under which various pairs of numbers have the same greatest common divisors. Notice in the next theorems that, although they look similar to the equation that we saw in the Division Algorithm, we use integers rather than natural numbers. Also, there is no hypothesis about the size of r in these theorems.

1.32 Theorem. *Let a, n, b, r, and k be integers. If $a = nb + r$ and $k|a$ and $k|b$, then $k|r$.*

1.33 Theorem. *Let a, b, n_1, and r_1 be integers with a and b not both 0. If $a = n_1b + r_1$, then $(a, b) = (b, r_1)$.*

1.34 Exercise. *As an illustration of the above theorem, note that*

$$51 = 3 \cdot 15 + 6,$$
$$15 = 2 \cdot 6 + 3,$$
$$6 = 2 \cdot 3 + 0.$$

Use the preceding theorem to show that if $a = 51$ and $b = 15$, then $(51, 15) = (6, 3) = 3$.

1.35 Exercise (Euclidean Algorithm). *Using the previous theorem and the Division Algorithm successively, devise a procedure for finding the greatest common divisor of two integers.*

The method you probably devised for finding the greatest common divisor of two integers is actually very famous. It dates back to the third century B.C. and is called the Euclidean Algorithm.

1.36 Exercise. *Use the Euclidean Algorithm to find*

 1. *(96, 112),*

 2. *(162, 31),*

 3. *(0, 256),*

 4. *(−288, −166),*

 5. *(1, −2436).*

The next exercise illustrates that the techniques that you are developing to find common divisors can also be used to find integer solutions to equations.

1.37 Exercise. *Find integers x and y such that $162x + 31y = 1$.*

This example is actually a special case of a general theorem that relates relatively prime numbers to integer solutions of equations.

Note: In the next theorem, remember as before that an "if and only if" theorem statement is really two separate theorems. As usual, to keep things clear, it's a good practice to write each down separately. We have done that for you again in this case to illustrate the practice.

Theorem. *Let a and b be integers. Then a and b are relatively prime (i.e., $(a, b) = 1$) if and only if there exist integers x and y such that $ax + by = 1$.*

Here, written separately, are the two theorems you must prove:

1.38 Theorem. *Let a and b be integers. If $(a, b) = 1$, then there exist integers x and y such that $ax + by = 1$.*

(Hint: Use the Euclidean Algorithm. Do some examples by taking some pairs of relatively prime integers, applying the Euclidean Algorithm, and seeing how to find the x and y. It is a good idea to start with an example where the Euclidean Algorithm takes just one step, then do an example where the Euclidean Algorithm takes two steps, then three steps, then look for a general procedure.)

1.39 Theorem. *Let a and b be integers. If there exist integers x and y with $ax + by = 1$, then $(a, b) = 1$.*

Once we have proved a theorem, we seek to find extensions or variations of it that are also true. In this case, we have just proved a theorem about relatively prime numbers. So it is natural to ask what we can say in the case that a pair of numbers is not relatively prime. We find that an analogous theorem is true.

1.40 Theorem. *For any integers a and b not both 0, there are integers x and y such that*

$$ax + by = (a, b).$$

The following three theorems appear here for two reasons; one, because you might use some of the previous results to prove them, and, two, because they will be useful for theorems to come.

1.41 Theorem. *Let a, b, and c be integers. If $a|bc$ and $(a, b) = 1$, then $a|c$.*

1.42 Theorem. *Let a, b, and n be integers. If $a|n$, $b|n$ and $(a, b) = 1$, then $ab|n$.*

1.43 Theorem. *Let a, b, and n be integers. If $(a, n) = 1$ and $(b, n) = 1$, then $(ab, n) = 1$.*

Our analysis so far of linear Diophantine equations will now prove to be quite useful in resolving our outstanding concern with cancellation in modular arithmetic. Recall your work in Question 1.20. Hopefully you showed the existence of integers a, b, c, and n (c not 0) for which $ac \equiv bc$ (mod n) and yet a is not congruent to b modulo n.

1.44 Question. *What hypotheses about a, b, c, and n could be added so that ac $\equiv$ bc (mod n) would imply a $\equiv$ b (mod n)? State an appropriate theorem and prove it before reading on.*

The next theorem answers the previous question, so be sure to answer Question 1.44 before reading further. The answer involves the concept of being relatively prime.

1.45 Theorem. *Let a, b, c and n be integers with n > 0. If ac $\equiv$ bc (mod n) and (c, n) = 1, then a $\equiv$ b (mod n).*

Theorems 1.39 and 1.40 begin to address the question: Given integers a, b, and c, when do there exist integers x and y that satisfy the equation $ax + by = c$? When we seek integer solutions to an equation, the equation is called a *Diophantine equation*.

1.46 Question. *Suppose a, b, and c are integers and that there is a solution to the linear Diophantine equation*

$$ax + by = c,$$

that is, suppose there are integers x and y that satisfy the equation ax + by = c. What condition must c satisfy in terms of a and b?

1.47 Question. *Can you make a conjecture by completing the following statement?*

Conjecture. *Given integers a, b, and c, there exist integers x and y that satisfy the equation ax + by = c if and only if _____.*

Try to prove your conjecture before reading further.

The following theorem summarizes the circumstances under which an equation $ax + by = c$ has integer solutions. It is an "if and only if" theorem, so, as always, you should write down the two separate theorems that must be proved.

1.48 Theorem. *Given integers a, b, and c with a and b not both 0, there exist integers x and y that satisfy the equation ax + by = c if and only if (a, b)|c.*

This theorem tells us under what conditions our linear equation has any solution; however, it does not tell us about all the solutions that such an equation might have, so it brings up a question.

1.49 Question. *For integers a, b, and c, consider the linear Diophantine equation*

$$ax + by = c.$$

Suppose integers x_0 and y_0 satisfy the equation; that is, $ax_0 + by_0 = c$. What other values

$$x = x_0 + h \text{ and } y = y_0 + k$$

also satisfy $ax + by = c$? Formulate a conjecture that answers this question. Devise some numerical examples to ground your exploration. For example, $6(-3) + 15 \cdot 2 = 12$. Can you find other integers x and y such that $6x + 15y = 12$? How many other pairs of integers x and y can you find? Can you find infinitely many other solutions?

The following question was devised by the famous mathematician Leonhard Euler (1707–1783). It presents a real life situation involving horses and oxen so that we can all identify with the problem. Can you see how Euler's problem is related to the preceding questions?

1.50 Exercise (Euler). *A farmer lays out the sum of $1,770$ crowns in purchasing horses and oxen. He pays 31 crowns for each horse and 21 crowns for each ox. What are the possible numbers of horses and oxen that the farmer bought?*

The following theorem shows you how to generate many solutions to our linear Diophantine equation, once you have one solution.

1.51 Theorem. *Let a, b, c, x_0, and y_0 be integers with a and b not both 0 such that $ax_0 + by_0 = c$. Then the integers*

$$x = x_0 + \frac{b}{(a,b)} \text{ and } y = y_0 - \frac{a}{(a,b)}$$

also satisfy the linear Diophantine equation $ax + by = c$.

This theorem leaves open the question of whether this method of generating alternative solutions generates all the solutions or whether there are yet more solutions.

1.52 Question. *If a, b, and c are integers with a and b not both 0, and the linear Diophantine equation*

$$ax + by = c$$

has at least one integer solution, can you find a general expression for all the integer solutions to that equation? Prove your conjecture.

The following theorem answers this question. It is actually two separate theorems that need two separate proofs. The first theorem says that certain numbers are solutions to $ax + by = c$. The second theorem, in the "Moreover" sentence, requires you to prove that no additional solutions exist.

1.53 Theorem. *Let a, b, and c be integers with a and b not both 0. If $x = x_0$, $y = y_0$ is an integer solution to the equation $ax + by = c$ (that is, $ax_0 + by_0 = c$) then for every integer k, the numbers*

$$x = x_0 + \frac{kb}{(a,b)} \text{ and } y = y_0 - \frac{ka}{(a,b)}$$

are integers that also satisfy the linear Diophantine equation $ax + by = c$. Moreover, every solution to the linear Diophantine equation $ax + by = c$ is of this form.

1.54 Exercise. *Find all integer solutions to the equation $24x + 9y = 33$.*

The previous theorem completes our analysis of the linear Diophantine equation

$$ax + by = c.$$

The analysis of the solutions of that Diophantine equation made good use of the greatest common divisor. We can now prove a theorem about greatest common divisors that might have been difficult to prove before we analyzed these Diophantine equations. However, it might be interesting to try to prove this simple sounding statement without using our theorems about Diophantine equations.

1.55 Theorem. *If a and b are integers, not both 0, and k is a natural number, then*

$$\gcd(ka, kb) = k \cdot \gcd(a, b).$$

We complete the chapter by taking the idea of greatest common divisor and considering a related idea. Common divisors of two numbers divide both numbers. A sort of opposite question is this: Suppose you are given two natural numbers. What numbers do those two numbers both divide; in other words, can we describe their common multiples? In particular, what is the least, common, positive multiple of two natural numbers? The first challenge is to write an appropriate definition.

1.56 Exercise. *For natural numbers a and b, give a suitable definition for "least common multiple of a and b", denoted* $\mathrm{lcm}(a,b)$. *Construct and compute some examples.*

The following theorem relates the ideas of the least common multiple and the greatest common divisor.

1.57 Theorem. *If a and b are natural numbers, then* $\gcd(a,b)\cdot\mathrm{lcm}(a,b) = ab$.

The next result is a corollary of Theorem 1.57. A corollary is a result whose proof follows directly from the statement of a previous theorem.

1.58 Corollary. *If a and b are natural numbers, then* $\mathrm{lcm}(a,b) = ab$ *if and only if a and b are relatively prime.*

After completing a body of work, it is satisfying and helpful to put together the ideas in your mind. We urge you to take that step by considering the following question.

1.59 Blank Paper Exercise. *After not looking at the material in this chapter for a day or two, take a blank piece of paper and outline the development of that material in as much detail as you can without referring to the text or to notes. Places where you get stuck or can't remember highlight areas that may call for further study.*

Linear Equations Through the Ages

Apart from introducing key concepts we will use throughout our investigations in number theory, we found in this chapter a complete solution to the linear Diophantine problem. What do we mean by "complete"? Given a linear equation $ax + by = c$ we can

1. determine whether or not the equation has integer solutions,

2. find an integer solution when one exists,

3. use a given solution to completely describe *all* integer solutions.

We will see in later chapters that such a degree of success in providing a complete solution to a Diophantine equation is not always so simple.

Problems of finding integer solutions to polynomial equations with integer coefficients have been dubbed Diophantine problems. Little is known of the Greek mathematician Diophantus of Alexandria. He most likely lived during the 3rd century A.D. (200–284 A.D.), and most of what survives from him today are six books from his treatise *Arithmetica*, a collection of 130 problems giving integer and rational solutions to equations. But unlike our results of this chapter, Diophantus was more concerned with particular problems and solutions rather than general methods.

General methods for finding solutions to linear Diophantine equations were first given by Indian mathematicians in the 5th century A.D. Notably, Aryabhata (476–550 A.D.), whose method of solving linear Diophantine equations translates as "pulverizer", and later, Brahmagupta (598–670 A.D.) described such procedures. For Aryabhata, the problem arose through the following consideration: can we find an integer n which when divided by a leaves a remainder r and when divided by b leaves a remainder r'? The problem's conditions can be translated into the following pair of equations

$$n = ax + r,$$
$$n = by + r'.$$

Equating the right-hand sides, and setting $c = r' - r$, gives the linear Diophantine equation

$$ax - by = c.$$

Progress did not occur in Western Europe for another 1000 years. It was not until the 17th century that their mathematicians began to piece together the solution as we have presented it in this chapter. Claude Bachet (1581–1638), most famous for his Latin translation of Diophantus' *Arithmetica*, rediscovered in 1621 a general method of finding a solution to $ax = by + 1$ when a and b are relatively prime. He employed a method much like ours, using the division algorithm repeatedly until a remainder of 1 is reached. Bachet then performed a sequence of "back substitutions" in a special way so as to avoid the need of negative numbers (which were not yet in common use).

Leonhard Euler may have been the first to give an actual proof that if a and b are relatively prime, then $ax + by = c$ is solvable in integers. What Euler in fact demonstrated is that the quantities $c - ka$, $k = 0, 1, \ldots, b - 1$ give b distinct remainders when divided by b. In particular, one, say $c - k'a$, yields a remainder of 0; that is, $c - k'a$ is equal to a multiple of b. Setting $c - k'a = nb$ then gives the solution $x = k'$ and $y = n$.

Joseph Lagrange (1736–1830), who also proved a version of Euler's result, went a step further to describe all integer solutions in terms of a given one. Perhaps he summed up the history of this problem best in stating that his method is "essentially the same as Bachet's, as are also all methods proposed by all mathematicians."

2

Prime Time

The Prime Numbers

One of the principal strategies by which we come to understand our physical or conceptual world is to break things down into pieces, describe the most basic pieces, and then describe how those pieces are assembled to create the whole. Our goal is to understand the natural numbers, so here we adopt that reductionist strategy and think about breaking natural numbers into pieces.

We begin by thinking about how natural numbers can be combined to create other natural numbers. The most basic method is through addition. So let's think about breaking natural numbers into their most basic pieces from the point of view of addition. What are the 'elements' so to speak with respect to addition of natural numbers? The answer is that there is only one element, the number 1. Every other natural number can be further broken down into smaller natural numbers that add together to create the number we started with. Every natural number is simply the sum of $1 + 1 + 1 + \cdots + 1$. Of course, this insight isn't too illuminating since every natural number looks very much like any other from this point of view. However, it does underscore the most basic property of the natural numbers, namely, that they all arise from the process of just adding 1 some number of times. In fact, this property of natural numbers lies at the heart of inductive processes both for constructing the natural numbers and often for proving theorems about them.

A more interesting way of constructing larger natural numbers from smaller ones is to use multiplication. Let's think about what the elementary particles, so to speak, are of the natural numbers with respect to multiplication. That is, what are the natural numbers that cannot be broken down into smaller natural numbers through multiplication. What natural numbers are not the product of smaller natural numbers? The answer, of course, is the prime numbers.

The study of primes is one of the main focuses of number theory. As we shall prove, every natural number greater than 1 is either prime or it can be expressed as a product of primes. Primes are the multiplicative building blocks of all the natural numbers.

The prime numbers give us a world of questions to explore. People have been exploring primes for literally thousands of years, and many questions about primes are still unanswered. We will prove that there are infinitely many primes, but how are they distributed among the natural numbers? How many primes are there less than a natural number n? How can we find them? How can we use them? These questions and others have been among the driving questions of number theory for centuries and have led to an incredible amount of beautiful mathematics.

New concepts in mathematics open frontiers of new questions and uncharted paths of inquiry. When we think of an idea, like the idea of prime numbers, we can pose questions about them to integrate the new idea with our already established web of knowledge. New mathematical concepts then arise by making observations, seeing connections, clarifying our ideas by making definitions, and then making generalizations or abstractions of what we have observed.

When we have isolated a concept sufficiently to make a definition, then we can state new theorems. We will see not only new theorems, but also new types of proof.

All proofs are simply sequences of statements that follow logically from one another, but one structure of proof that you will develop and use in this chapter and future chapters is proof by induction. You will naturally come up with inductive styles of proving theorems on your own. In fact you may already have used this kind of argument in the last chapter, for example, in proving that the Euclidean Algorithm works. Inductive styles of proof are so useful that it is worthwhile to reflect on the logic involved. We have included an appendix that describes this technique of proof, and this may be a good time to work through that appendix.

Fundamental Theorem of Arithmetic

The role of definitions in mathematics cannot be overemphasized. They allow us to be precise in our language and reasoning. When a new definition is introduced, you should take some time to familiarize yourself with its details. Try to get comfortable with its meaning. Look at examples. Memorize it.

Definition. A natural number $p > 1$ is *prime* if and only if p is not the product of natural numbers less than p.

Definition. A natural number n is *composite* if and only if n is a product of natural numbers less than n.

The following theorem tells us that every natural number larger than 1 has at least one prime factor.

2.1 Theorem. *If n is a natural number greater than 1, then there exists a prime p such that $p|n$.*

To get accustomed to primes, it's a good idea to find some.

2.2 Exercise. *Write down the primes less than 100 without the aid of a calculator or a table of primes and think about how you decide whether each number you select is prime or not.*

You probably identified the primes in the previous exercise by trial division. For example, to determine whether or not 91 was prime, you might have first tried dividing it by 2. Once convinced that 2 does not divide 91, you probably moved on to 3; then 4; then 5; then 6. Finally, you reached 7 and discovered that in fact 91 is not a prime. You were probably relieved, as you might have secretly feared that you would have to continue the daunting task of trial division 91 times! The following theorem tells us that you need not have been too concerned.

2.3 Theorem. *A natural number $n > 1$ is prime if and only if for all primes $p \leq \sqrt{n}$, p does not divide n.*

2.4 Exercise. *Use the preceding theorem to verify that 101 is prime.*

The search for prime numbers has a long and fascinating history that continues to unfold today. Recently the search for primes has taken on practical significance because primes are used everyday in making internet communications secure, for example. Later, we will investigate ways that primes are used in cryptography. And we'll see some modern techniques of identifying primes. But let's begin with an ancient method for finding primes. The following exercise introduces a very early method of identifying primes attributed to the scholar Eratosthenes (276–194 BC).

2.5 Exercise (Sieve of Eratosthenes). *Write down all the natural numbers from 1 to 100, perhaps on a 10×10 array. Circle the number 2, the smallest prime. Cross off all numbers divisible by 2. Circle 3, the next number that is not crossed out. Cross off all larger numbers that are divisible by 3. Continue to circle the smallest number that is not crossed out and cross out its multiples. Repeat. Why are the circled numbers all the primes less than 100?*

With our list of primes, we can begin to investigate how many primes there are and what proportion of natural numbers are prime.

2.6 Exercise. *For each natural number n, define $\pi(n)$ to be the number of primes less than or equal to n.*

1. *Graph $\pi(n)$ for $n = 1, 2, \ldots, 100$.*

2. *Make a guess about approximately how large $\pi(n)$ is relative to n. In particular, do you suspect that $\frac{\pi(n)}{n}$ is generally an increasing function or a decreasing function? Do you suspect that it approaches some specific number (as a limit) as n goes to infinity? Make a conjecture and try to prove it. Proving your conjecture is a difficult challenge. You might use a computer to extend your list of primes to a much larger number and see whether your conjecture seems to be holding up.*

Mathematicians do not give out the title of "Fundamental Theorem" too often. In fact, you may have only come across one or two in your lifetime (the Fundamental Theorem of Algebra and the Fundamental Theorem of Calculus come to mind). We might think of such theorems as somehow very important. If so, we would be correct. What makes a theorem important? One answer might be that it captures a basic relationship and that it is widely applicable to explaining a broad range of mathematics. We will see that the Fundamental Theorem of Arithmetic certainly possesses these qualities.

We will write the Fundamental Theorem of Arithmetic in two parts: the Existence part and the Uniqueness part. The Existence part says that every natural number bigger than 1 can be written as the product of primes and the Uniqueness part says basically that there is only one way to do so. For example, $24 = 2^3 \cdot 3 = 3 \cdot 2^3$.

2.7 Theorem (Fundamental Theorem of Arithmetic—Existence Part). *Every natural number greater than 1 is either a prime number or it can be expressed as a finite product of prime numbers. That is, for every natural number n greater than 1, there exist distinct primes $p_1, p_2, \ldots, p_m$ and natural numbers $r_1, r_2, \ldots, r_m$ such that*

$$n = p_1^{r_1} p_2^{r_2} \cdots p_m^{r_m}.$$

The following lemma might be helpful in proving the Uniqueness part of the Fundamental Theorem of Arithmetic. A lemma is actually a theorem, but it is designed to be a step towards the proof of a more important theorem.

2.8 Lemma. *Let p and $q_1, q_2, \ldots, q_n$ all be primes and let k be a natural number such that $pk = q_1 q_2 \cdots q_n$. Then $p = q_i$ for some i.*

2.9 Theorem (Fundamental Theorem of Arithmetic—Uniqueness part). *Let n be a natural number. Let $\{p_1, p_2, \ldots, p_m\}$ and $\{q_1, q_2, \ldots, q_s\}$ be sets of primes with $p_i \neq p_j$ if $i \neq j$ and $q_i \neq q_j$ if $i \neq j$. Let $\{r_1, r_2, \ldots, r_m\}$ and $\{t_1, t_2, \ldots, t_s\}$ be sets of natural numbers such that*

$$n = p_1^{r_1} p_2^{r_2} \cdots p_m^{r_m}$$
$$= q_1^{t_1} q_2^{t_2} \cdots q_s^{t_s}.$$

Then $m = s$ and $\{p_1, p_2, \ldots, p_m\} = \{q_1, q_2, \ldots, q_s\}$. That is, the sets of primes are equal but their elements are not necessarily listed in the same order; that is, p_i may or may not equal q_i. Moreover, if $p_i = q_j$ then $r_i = t_j$. In other words, if we express the same natural number as a product of powers of distinct primes, then the expressions are identical except for the ordering of the factors.

Putting the existence and uniqueness parts together, we get the whole formulation of the Fundamental Theorem of Arithmetic:

Theorem (Fundamental Theorem of Arithmetic). *Every natural number greater than 1 is either a prime number or it can be expressed as a finite product of prime numbers where the expression is unique up to the order of the factors.*

Let's take a moment to think through a little issue about our definition of "prime." Humans make decisions about what definitions to make. Let's think about the choices we made in defining "prime." One notion of "prime" is the inability to further decompose. Surely 1 meets this criterion. Yet our choice of definition of prime omits 1. What is the advantage to not choosing to include 1 among the prime numbers? If 1 were called a prime, why would the Fundamental Theorem of Arithmetic no longer be true?

The Fundamental Theorem of Arithmetic tells us that every natural number bigger than 1 is a product of primes. Here are some exercises that help to show what that means in some specific cases.

2.10 Exercise. *Express* $n = 12!$ *as a product of primes.*

2.11 Exercise. *Determine the number of zeroes at the end of* $25!$.

The Fundamental Theorem of Arithmetic says that for any natural number $n > 1$ there exist distinct primes $\{p_1, p_2, \ldots, p_m\}$ and natural numbers $\{r_1, r_2, \ldots, r_m\}$ such that

$$n = p_1^{r_1} p_2^{r_2} \cdots p_m^{r_m}$$

and moreover, the factorization is unique up to order. When the p_i are ordered so that $p_1 < p_2 < \cdots < p_m$ we will say that $p_1^{r_1} p_2^{r_2} \cdots p_m^{r_m}$ is the *unique prime factorization of* n.

Applications of the Fundamental Theorem of Arithmetic

One application of the Fundamental Theorem of Arithmetic is that if we know the prime factorizations of two natural numbers, it is a simple matter to determine whether one divides the other. The following is a characterization of divisibility in terms of primes. There are lots of letters and lots of subscripts, but once understood, this theorem makes sense.

2.12 Theorem. *Let a and b be natural numbers greater than 1 and let $p_1^{r_1} p_2^{r_2} \cdots p_m^{r_m}$ be the unique prime factorization of a and let $q_1^{t_1} q_2^{t_2} \cdots q_s^{t_s}$ be the unique prime factorization of b. Then $a|b$ if and only if for all $i \leq m$ there exists a $j \leq s$ such that $p_i = q_j$ and $r_i \leq t_j$.*

Prime factorizations make it easy to prove some assertions that might otherwise be more difficult.

2.13 Theorem. *If a and b are natural numbers and $a^2|b^2$, then $a|b$.*

Prime factorizations can help us to find the greatest common divisor and least common multiple of two natural numbers. Here are some examples.

2.14 Exercise. *Find $(3^{14} \cdot 7^{22} \cdot 11^5 \cdot 17^3, 5^2 \cdot 11^4 \cdot 13^8 \cdot 17)$.*

2.15 Exercise. *Find* $\text{lcm}(3^{14} \cdot 7^{22} \cdot 11^5 \cdot 17^3, 5^2 \cdot 11^4 \cdot 13^8 \cdot 17)$.

After doing some examples, we instinctively seek the general pattern. That is, we seek to make a general statement that captures the reason that the method we used in the specific examples works.

2.16 Exercise. *Make a conjecture that generalizes the ideas you used to solve the two previous exercises.*

2.17 Question. *Do you think this method is always better, always worse, or sometimes better and sometimes worse than using the Euclidean Algorithm to find (a, b)? Why?*

The following theorem requires a clever use of the Fundamental Theorem of Arithmetic.

2.18 Theorem. *Given $n + 1$ natural numbers, say $a_1, a_2, \ldots, a_{n+1}$, all less than or equal to $2n$, then there exists a pair, say a_i and a_j with $i \neq j$, such that $a_i | a_j$.*

The Fundamental Theorem of Arithmetic can be used to prove that certain equations do not have integer solutions.

2.19 Theorem. *There do not exist natural numbers m and n such that $7m^2 = n^2$.*

2.20 Theorem. *There do not exist natural numbers m and n such that $24m^3 = n^3$.*

Up to this point we have been talking exclusively about natural numbers and integers. Our insights into natural numbers and integers can actually help us to understand more general kinds of numbers such as rational numbers and irrational numbers.

Definition. A *rational number* is a real number that can be written as $\frac{a}{b}$ where a and b are integers and $b \neq 0$.

Definition. A real number that is not rational is *irrational*.

The next theorems ask you to prove that certain specific numbers are irrational.

2.21 Exercise. *Show that $\sqrt{7}$ is irrational. That is, there do not exist natural numbers n and m such that $\sqrt{7} = \frac{n}{m}$.*

2.22 Exercise. *Show that* $\sqrt{12}$ *is irrational.*

2.23 Exercise. *Show that* $7^{\frac{1}{3}}$ *is irrational.*

Having proved some specific numbers are irrational we take the usual step of generalizing our insights as far as possible.

2.24 Question. *What other numbers can you show to be irrational? Make and prove the most general conjecture you can.*

Let's now return to the world of integers. The following was a theorem we first proved in Chapter 1. Here we repeat the theorem with the idea that the Fundamental Theorem of Arithmetic might help to provide an alternative proof.

2.25 Theorem. *Let a, b, and n be integers. If $a|n$, $b|n$, and $(a,b) = 1$, then $ab|n$.*

Integers are either divisible by a prime p or are relatively prime to p.

2.26 Theorem. *Let p be a prime and let a be an integer. Then p does not divide a if and only if $(a,p) = 1$.*

Notice that $9|(6 \cdot 12)$ and yet 9 does not divide either 6 or 12. However, if a prime divides a product of two integers, then it must divide one or the other.

2.27 Theorem. *Let p be a prime and let a and b be integers. If $p|ab$, then $p|a$ or $p|b$.*

The following theorems explore the relationships among the greatest common divisor and various arithmetic operations. You might consider proving them in at least two ways, one using the Fundamental Theorem of Arithmetic and one using the techniques from Chapter 1.

2.28 Theorem. *Let a, b, and c be integers. If $(b,c) = 1$, then $(a,bc) = (a,b) \cdot (a,c)$.*

2.29 Theorem. *Let a, b, and c be integers. If $(a,b) = 1$ and $(a,c) = 1$, then $(a,bc) = 1$.*

2.30 Theorem. *Let a and b be integers. If $(a,b) = d$, then $(\frac{a}{d}, \frac{b}{d}) = 1$.*

2.31 Theorem. *Let a, b, u, and v be integers. If* $(a, b) = 1$ *and* $u|a$ *and* $v|b$, *then* $(u, v) = 1$.

The infinitude of primes

One of the most basic questions we can ask about prime numbers is, "How many are there?" In this section, we will prove that there are infinitely many. To prove that there are infinitely many primes, we need to show that there are large natural numbers that are not the product of smaller natural numbers. Our first observation points out that consecutive natural numbers cannot share common divisors greater than 1.

2.32 Theorem. *For all natural numbers n,* $(n, n + 1) = 1$.

Can you think of a natural number that is divisible by 2, 3, 4, and 5? Can you think of a natural number that has a remainder of 1 when divided by 2, 3, 4, and 5? If you think of systematic ways to answer these questions, you will be well on your way to proving the following theorem.

2.33 Theorem. *Let k be a natural number. Then there exists a natural number n (which will be much larger than k) such that no natural number less than k and greater than* 1 *divides n.*

The previous theorem shows us how to produce natural numbers that are specifically not divisible by certain natural numbers. This insight helps us to find natural numbers that are not divisible by any natural numbers other than themselves and 1, in other words, primes.

2.34 Theorem. *Let k be a natural number. Then there exists a prime larger than* k.

The Infinitude of Primes Theorem is one of the basic results of mathematics. It was proved in ancient times and is recognized as one of the foundational theorems about numbers. At first you might think, "Of course, there must be infinitely many primes. How could there not be infinitely many primes since there are infinitely many natural numbers?" But remember that the same prime can be used many times. For example, we can construct arbitrarily large natural numbers just by raising 2 to large powers. So it is conceivable that some finite number of primes would suffice to produce all natural numbers. However, in fact there are infinitely many primes, as you will now prove.

2.35 Theorem (Infinitude of Primes Theorem). *There are infinitely many prime numbers.*

After you have devised a proof or proofs or learned a proof, it is satisfying to reflect on the logic of the argument and celebrate and appreciate the beauty or cleverness of the reasoning.

2.36 Question. *What were the most clever or most difficult parts in your proof of the Infinitude of Primes Theorem?*

One of the principal ways that new mathematics is created is to take one result and see whether it can be extended or variations of it can be proved. In the case of the Infinitude of Primes, we can ask whether there are infinitely many primes of a certain type. We begin by making an observation about numbers congruent to 1 modulo 4, which then will help us to prove that there are infinitely many primes of the form $4k + 3$.

2.37 Theorem. *If $r_1, r_2, \ldots, r_m$ are natural numbers and each one is congruent to 1 modulo 4, then the product $r_1 r_2 \cdots r_m$ is also congruent to 1 modulo 4.*

To prove the following theorem, remember the proof of the Infinitude of Primes Theorem and see how the strategy of that proof might be adapted to prove the following harder theorem.

2.38 Theorem (Infinitude of $4k + 3$ Primes Theorem). *There are infinitely many prime numbers that are congruent to 3 modulo 4.*

When you have proved the previous theorem, you will have forced yourself to understand a technique of proving theorems about the existence of infinitely many primes of a certain type. Now is the time to see how far that technique can be pushed. In other words ask yourself how many theorems like the preceding one are provable using a similar idea.

2.39 Question. *Are there other theorems like the previous one that you can prove?*

In fact, the following much more general theorem is true. Its proof in its full generality, however, is quite difficult and we will not attempt it in this course.

Theorem (Infinitude of $ak + b$ Primes Theorem). *If a and b are relatively*

*prime natural numbers, then there are infinitely many natural numbers k
for which $ak + b$ is prime.*

The previous theorem is often called *Dirichlet's Theorem on primes in
an arithmetic progression* and is due to Lejeune Dirichlet (1805–1859).
An arithmetic progression is a sequence of numbers of the form $ak + b$,
$k = 0, 1, 2, \ldots$, where b is any integer and a is a natural number. It is a
sequence of numbers all of which are congruent to b modulo a. The study
of primes in arithmetic progressions is still an active field today. Consider
the following recent result due to Ben Green and Terence Tao.

Theorem (Green and Tao, 2005). *There are arbitrarily long arithmetic
progressions of primes.*

This means that for any natural number n there exists a prime p and a
natural number a such that p, $p + a$, $p + 2a$, $p + 3a$, $\ldots$, $p + na$ are
all prime. As an example, an arithmetic progression of primes of length
five is found by choosing $p = 5$ and $a = 6$, which yields the sequence
$5, 11, 17, 23, 29$. The longest known arithmetic progression of primes as of
July of 2004 has length 23 and is given by

$$56211383760397 + k\,44546738095860, \; k = 0, \ldots, 22.$$

Terence Tao was awarded a Fields medal in part for his work related to
this result. Fields medals, the mathematical equivalent of the Nobel prize,
are awarded once every four years to outstanding mathematicians under the
age of 40.

2.40 Exercise. *Find the current record for the longest arithmetic progression of primes.*

Primes of special form

The largest known prime is of a special type known as a Mersenne prime,
which is a prime of the form $2^n - 1$. The theorems here show some features
of Mersenne primes and related primes.

2.41 Exercise. *Use polynomial long division to compute $(x^m - 1) \div (x - 1)$.*

2.42 Theorem. *If n is a natural number and $2^n - 1$ is prime, then n must
be prime.*

2.43 Theorem. *If n is a natural number and $2^n + 1$ is prime, then n must be a power of 2.*

Definition. A *Mersenne prime* is a prime of the form $2^p - 1$, where p is a prime. A prime of the form $2^{2^k} + 1$ is called a *Fermat prime.*

2.44 Exercise. *Find the first few Mersenne primes and Fermat primes.*

2.45 Exercise. *For an A in the class and a Ph.D. in mathematics, prove that there are infinitely many Mersenne primes (or Fermat primes) or prove that there aren't (your choice).*

The distribution of primes

We now know that there are infinitely many primes, but in a sense that information is a rather crude measure of how the primes appear among the natural numbers. We could ask other questions such as roughly what fraction of the natural numbers are prime? And we might wonder whether the primes occur in some sort of pattern. To investigate how the primes are distributed among the natural numbers, let's begin by looking at some ranges of natural numbers with the primes printed in bold:

$$1, \mathbf{2}, \mathbf{3}, 4, \mathbf{5}, 6, \mathbf{7}, 8, 9, 10, \mathbf{11}, 12, \mathbf{13}, 14, 15, 16, \mathbf{17}, 18, \mathbf{19}, 20, 21,$$

$$22, \mathbf{23}, 24, \dots, 300, 301, 302, 303, 304, 305, 306, \mathbf{307}, 308, 309,$$

$$310, \mathbf{311}, 312, \mathbf{313}, 314, 315, 316, \dots, 2025, 2026, \mathbf{2027}, 2028,$$

$$\mathbf{2029}, 2030, 2031, 2032, 2033, 2034, 2035, 2036, 2037, 2038, \dots$$

What observations can we make? First, we may notice that the proportion of bold numbers occurring seems to be getting smaller. That is, primes tend to be more sparse as we move further out into the sequence of natural numbers. We tend to see longer and longer runs of consecutive composite numbers. In fact, there is no limit to the length of strings of composite numbers.

2.46 Theorem. *There exist arbitrarily long strings of consecutive composite numbers. That is, for any natural number n there is a string of more than n consecutive composite numbers.*

On the other hand, we still observe pairs of primes separated by just one even number, such as 311 and 313, or 2027 and 2029. One of the most famous unanswered questions in number theory asks whether or not this behavior continues indefinitely. If you have already settled the previous question about Mersenne primes, then solving the following question will give you another Ph.D.

2.47 Question (The Twin Primes Question). *Are there infinitely many pairs of prime numbers that differ from one another by two? (The pairs* 11 *and* 13, 29 *and* 31, 41 *and* 43 *are examples of such twin primes.)*

Out of the first 24 natural numbers, 9 of them are primes—that's just a little over one third. We saw how this fraction changes as n increases in the Sieve of Eratosthenes exercise.

Suppose someone asked you to write down all the primes less than 100 million without the aid of a calculator or a computer. With a pencil and paper, you would find that task to be tedious and prone to error; however, that was the challenge facing mathematicians before the advent of modern computing machinery. Surely one of the most amazing feats of prime-finding before computers was completed in about 1863, when J.P. Kulik finished his 20-year project of finding the least prime factor of every natural number up to 100 million. Our sadness in losing the volume of Kulik's work that contained the natural numbers between 12,642,600 and 22,852,800 is somewhat lessened by the fact that his work was full of errors and that a modern computer could reproduce the whole work in a matter of seconds.

The significance of computing lists of primes before the invention of computers and even before Kulik's work is that those lists allowed mathematicians to gain some intuition about the distribution of primes.

As we observed above, the *proportion* of primes seems to slowly go downward. That is, the *percentage* of numbers less than a million that are prime is smaller than the *percentage* of numbers less than a thousand that are prime. The primes, in some sense, get sparser and sparser among the bigger numbers. That observation was greatly refined in the 1790s by Carl Friedrich Gauss (1777–1855), known by many as the Prince of Mathematics, and Adrien-Marie Legendre (1752–1833). They conjectured that the number of primes less than the natural number n, which is denoted by $\pi(n)$, is approximated by n divided by the *natural logarithm* of n. Using computers, we can produce evidence that the proportion of primes less than n becomes increasingly smaller as n increases. Table 1 also shows that the ratio between $\pi(n)$ and the fraction $\frac{n}{\ln(n)}$ gets increasingly closer to 1.

n	$\pi(n)$	$\frac{\pi(n)}{n}$	$\frac{n}{\ln(n)}$	$\frac{\pi(n)}{n/\ln(n)}$
10	4	.4	4.3...	0.92104...
10^2	25	.25	21.7...	1.15133...
10^3	168	.168	144.7...	1.16054...
10^4	1229	.1229	1085.7...	1.13199...
10^5	9592	.09592	8685.8...	1.10443...
10^6	78498	.078498	72382.4...	1.08452...
10^7	664579	.0664579	620420.7...	1.07121...
10^8	5761455	.05761455	5428681.0...	1.06144...
10^9	50847534	.050847534	48254942.4...	1.05385...

Table 1. Prime Proportions

The formal statement of these observations is called The Prime Number Theorem. We state it here, but the proofs of this theorem are difficult, and beyond the scope of this book.

Theorem (The Prime Number Theorem). *As n approaches infinity, the number of primes less than n, $\pi(n)$, approaches $\frac{n}{\ln(n)}$, that is,*

$$\lim_{n\to\infty}\left(\frac{\pi(n)}{n/\ln(n)}\right)=1.$$

Finally, we mention here one more famous open question concerning prime numbers.

2.48 Exercise. *Express each of the first 20 even numbers greater than 2 as a sum of two primes. (For example: $8 = 5 + 3$.)*

In a letter to Euler, dated June 7, 1742, Christian Goldbach (1690–1764) claimed that every natural number greater than 2 was the sum of three primes. It was convention at the time to include the number 1 as being among the primes. The conjecture was re-expressed by Euler as follows.

Conjecture (The Goldbach Conjecture). *Every positive, even number greater than 2 can be written as the sum of two primes.*

The Goldbach Conjecture has been verified by computer, as of June of 2006, for all even numbers up to 400, 000, 000, 000, 000, 000. As the even numbers get larger, there seem to be more ways to write them as a sum of two primes. For example, the number 100,000,000 can be written as the sum of two primes in 219,400 different ways. But no one knows how to prove that in general all even natural numbers are the sum of two primes. Perhaps some even number with 10 trillion digits is not the sum of two primes. Until we have a general method of proof that will apply to all even numbers, we will not know whether such a natural number exists or not.

2.49 Blank Paper Exercise. *After not looking at the material in this chapter for a day or two, take a blank piece of paper and outline the development of that material in as much detail as you can without referring to the text or to notes. Places where you get stuck or can't remember highlight areas that may call for further study.*

From Antiquity to the Internet

Interest in the multiplicative properties of the natural numbers surely predated the works of Euclid (*Elements*, Books VII, VIII, IX), but it is here that we find the first written study. For example, Proposition 20 of Book IX gives the first known proof of the infinitude of primes. The ancient Greeks' interest in the primes may have been further spawned by the connection they shared with *perfect numbers*. A natural number is said to be *perfect* if it is equal to the sum of its proper divisors. For example, the smallest perfect number is 6, since $6=1+2+3$. We list the first four perfect numbers.

$$6 = 2^{2-1}(2^2 - 1) = 1 + 2 + 3$$
$$28 = 2^{3-1}(2^3 - 1) = 1 + 2 + 4 + 7 + 14$$
$$496 = 2^{5-1}(2^5 - 1) = 1 + 2 + 4 + 8 + 16 + 31 + 62 + 124 + 248$$
$$8128 = 2^{7-1}(2^7 - 1) = 1 + 2 + 4 + 8 + 16 + \cdots + 2032 + 4064$$

In Book IX of his *Elements* Euclid proved the following: if for some n, $2^n - 1$ is prime, then $2^{n-1}(2^n - 1)$ is perfect. This established the link between perfect numbers and primes of the form $2^n - 1$.

The serious study of perfect numbers and primes of special forms was picked up again in the seventeenth century by the likes of Rene Descartes (1596–1650), Pierre de Fermat (1601–1665), and Marin Mersenne (1588–1648). In a 1638 letter to Mersenne, Descartes stated that he thought he could prove that every even perfect number was of the form given by Euclid's theorem, but no proof was given. Also in a letter to Mersenne, dated 1640, Fermat indicated he had proved the following: if n is composite, then $2^n - 1$ is composite; but if n is prime, then $2^n - 1$ need not be prime, with two examples being $2^{11} - 1 = 23 \cdot 89$, and $2^{23} - 1 = 47 \cdot 178481$.

In 1647 Mersenne gave the following list of 11 primes p for which he believed $2^p - 1$ was prime as well: 2, 3, 5, 7, 13, 17, 19, 31, 67, 127, 257. He erred only by including 67 (and excluding 61, 89 and 107). To this day primes of the form $2^p - 1$ are called Mersenne primes, and it is still unknown whether infinitely many exist. In a posthumously published paper, Euler finally succeeded in proving that all even perfect numbers are of Euclid's type, giving a one-to-one correspondence between Mersenne primes and even perfect numbers. Curiously, it is not known if any odd perfect numbers exist.

The search for new Mersenne primes continues to this day. In fact, anyone with a home computer and an internet connection can join the *Great Internet Mersenne Prime Search* (GIMPS). Mersenne's list has only been increased to contain 44 examples as of September, 2006, with the largest having over 9.8 million digits.

2.50 Exercise. *Find the current record for the largest known Mersenne prime.*

There is a monetary award of $100,000 for the first person (or group) to find a Mersenne prime with at least 10 million digits. So happy hunting.

3

A Modular World

Thinking Cyclically

In Chapter 1 we established the basics of modular arithmetic. Now we proceed to see how modular arithmetic relates to other familiar algebraic constructions such as functions and equations, and how it can help us to better understand primes and composite numbers.

Modular arithmetic is interesting as an abstract topic in number theory, but it also plays important roles in real life. It is the basis for public key cryptography and check digits associated with error detection. Here we further develop the theory of modular arithmetic and later explore some of its applications outside mathematics.

Powers and polynomials modulo n

Recall the following definition of *congruence* from Chapter 1.

Definition. Suppose that a, b, and n are integers with $n > 0$. We say that a and b are *congruent modulo n* if and only if $n|(a - b)$. We denote this relationship as

$$a \equiv b \pmod{n}$$

and read these symbols as "a is congruent to b modulo n."

Here are some exercises that will encourage you to refresh your memory about some of the modular arithmetic theorems that you proved back in Chapter 1.

3.1 Exercise. *Show that* 41 *divides* $2^{20} - 1$ *by following these steps. Explain why each step is true.*

1. $2^5 \equiv -9 \pmod{41}$.

2. $(2^5)^4 \equiv (-9)^4 \pmod{41}$.

3. $2^{20} \equiv 81^2 \pmod{41} \equiv (-1)^2 \pmod{41}$.

4. $2^{20} - 1 \equiv 0 \pmod{41}$.

3.2 Question. *In your head, can you find the natural number* k, $0 \le k \le$ 11, *such that* $k \equiv 37^{453} \pmod{12}$?

(Hint: Don't try to multiply it out and then divide by 12. *Of course, this hint is a rather lame joke, since if you could actually multiply* 37^{453} *in your head, you would not be taking a number theory class. You would be performing mental feats in some carnival sideshow.)*

The next question continues to show you the value of thought (and modular arithmetic) rather than brute force.

3.3 Question. *In your head or using paper and pencil, but no calculator, can you find the natural number* k, $0 \le k \le 6$, *such that* $2^{50} \equiv k \pmod 7$?

The next question asks you to compute a larger power (453) of a number modulo 12. Try to think of how to do this efficiently. Here is a hint. If you want to raise a number to the 16th power, you can first square it, then square the result, then square the result, and then square the result. So only four multiplications accomplish raising to the 16th power, rather than using 16 multiplications. Also, remember that you can reduce answers modulo 12, so you never have to multiply numbers larger than 11. While doing the following exercise, think about systematizing your strategy. In particular, can you see why your strategy might involve expressing 453 as a sum of powers of 2? See whether you can do the following problem without ever multiplying numbers larger than 12 and without doing more than 10 steps of multiplying two numbers less than 12 and reducing the answers modulo 12.

3.4 Question. *Using paper and pencil, but no calculator, can you find the natural number* k, $0 \le k \le 11$, *such that* $39^{453} \equiv k \pmod{12}$?

Now that you have developed the power to take powers, here is another exercise that takes advantage of your method.

3.5 Exercise. *Show that* 39 *divides* $17^{48} - 5^{24}$.

At this point, you have developed some ideas about how to efficiently raise numbers to powers in modular arithmetic. The next question asks you to crystallize your method and clearly describe it.

3.6 Question (Describe technique). *Let a, n, and r be natural numbers. Describe how to find the number k $(0 \leq k \leq n - 1)$ such that $k \equiv a^r$ (mod n) subject to the restraint that you never multiply numbers larger than n and that you only have to do about $\log_2 r$ such multiplications.*

The technique you just developed and described allows computers to deal with taking very large numbers (containing several hundred digits) and raising them to huge powers modulo other enormous numbers. The ability of computers to deal with such arithmetical challenges turns out to be an essential ingredient in modern methods of secure data transmission used over the internet everyday. We will explore these methods, which involve cryptography, in a later chapter.

We now turn our attention to polynomials and how they behave when viewed from a modular arithmetic point of view. We begin with a specific example.

3.7 Question. *Let $f(x) = 13x^{49} - 27x^{27} + x^{14} - 6$. Is it true that*

$$f(98) \equiv f(-100) \quad (\text{mod } 99)?$$

As usual, after doing a specific example, we think about what more general statement the specific example suggests.

3.8 Theorem. *Suppose $f(x) = a_n x^n + a_{n-1} x^{n-1} + \cdots + a_0$ is a polynomial of degree $n > 0$ with integer coefficients. Let a, b, and m be integers with $m > 0$. If $a \equiv b$ (mod m), then $f(a) \equiv f(b)$ (mod m).*

The next corollaries are repeats of results from Chapter 1 about criteria for determining when a natural number is divisible by 3 or 9. Here you are being asked to recognize a natural number as the evaluation of a polynomial, and to deduce the subsequent statements from the previous theorem.

3.9 Corollary. *Let the natural number n be expressed in base 10 as*

$$n = a_k a_{k-1} \ldots a_1 a_0.$$

Let $m = a_k + a_{k-1} + \cdots + a_1 + a_0$. Then $9|n$ if and only if $9|m$.

3.10 Corollary. *Let the natural number n be expressed in base 10 as*

$$n = a_k a_{k-1} \ldots a_1 a_0.$$

If $m = a_k + a_{k-1} + \cdots + a_1 + a_0$. *Then* $3|n$ *if and only if* $3|m$.

During your work on Chapter 1, you may have devised other criteria for divisibility. If so, does this polynomial view of those divisibility theorems help you to see why your methods are true? Can you now think of new divisibility theorems like the above?

The next two theorems do not involve modular arithmetic. They roughly state that every polynomial gets big.

3.11 Theorem. *Suppose* $f(x) = a_n x^n + a_{n-1} x^{n-1} + \cdots + a_0$ *is a polynomial of degree* $n > 0$ *and suppose* $a_n > 0$. *Then there is an integer k such that if* $x > k$, *then* $f(x) > 0$.

Note: We are only assuming that the leading coefficient a_n is greater than zero. The other coefficients may be positive or negative or zero.

The next theorem extends the idea that polynomials get positive and roughly states that not only do they get positive, but they get big and stay big from some point on. Notice that the theorem does not ask you to be efficient and find the first place after which the polynomial stays larger than some value. It just asks you to prove that eventually that happens.

3.12 Theorem. *Suppose* $f(x) = a_n x^n + a_{n-1} x^{n-1} + \cdots + a_0$ *is a polynomial of degree* $n > 0$ *and suppose* $a_n > 0$. *Then for any number M there is an integer k (which depends on M) such that if* $x > k$, *then* $f(x) > M$.

The next theorem connects polynomials with primes. It says that every polynomial with integer coefficients produces many composite numbers. There is no polynomial that produces only primes. Too bad. In proving the next theorem, it might be useful to think about modular arithmetic. Remember that if a number is congruent to 0 modulo n, then n divides the number, and being divisible is the fundamental issue about being composite. The proof of the following theorem is a challenge, but if you look at it just right, then you can give a convincing proof. So the hint is to use Theorems 3.8 and 3.12.

3.13 Theorem. *Suppose* $f(x) = a_n x^n + a_{n-1} x^{n-1} + \cdots + a_0$ *is a polynomial of degree* $n > 0$ *with integer coefficients. Then* $f(x)$ *is a composite number for infinitely many integers* x.

Note: This theorem implies that we cannot find a magical polynomial that produces only prime values for every integer input. Nevertheless, some polynomials do pretty well. The polynomial $f(x) = x^2 + x + 41$ has a prime value (that is, $f(n)$ is prime) for 80 consecutive integer inputs, $n = -40, -39, \ldots, 38, 39$. Try a few values to test this assertion.

When we think of a natural number modulo n, it is congruent to some non-negative integer less than n. The next theorem pins that idea down.

3.14 Theorem. *Given any integer a and any natural number n, there exists a unique integer t in the set $\{0, 1, 2, \ldots, n - 1\}$ such that $a \equiv t \pmod{n}$.*

This theorem suggests the following definition of one set of numbers to which every natural number is congruent.

Definition. Let n be a natural number. The set $\{0, 1, 2, \ldots, n - 1\}$ is the called the *canonical complete residue system modulo n*.

There are other collections of integers besides the canonical complete residue system modulo n with the property that they represent all integers modulo n.

Definition. Let k and n be natural numbers. A set $\{a_1, a_2, \ldots, a_k\}$ of integers is called a *complete residue system modulo n* if every integer is congruent modulo n to exactly one element of the set.

Let's get used to these definitions by looking at some examples and constructing complete residue systems.

3.15 Exercise. *Find three complete residue systems modulo 4: the canonical complete residue system, one containing negative numbers, and one containing no two consecutive numbers.*

3.16 Theorem. *Let n be a natural number. Every complete residue system modulo n contains n elements.*

Arithmetic modulo n puts the integers into n different *equivalence classes*. A complete residue system modulo n has one representative of each equivalence class. Even if you don't know the technical definition of equivalence class, the idea is just that the integers are divided into groups, namely, the integers congruent to 0, the integers congruent to 1, the integers congruent to 2, and so on up to the integers congruent to $n - 1$ modulo n. The following theorem says that any set of n non-congruent integers will form a complete residue system modulo n.

3.17 Theorem. *Let n be a natural number. Any set, $\{a_1, a_2, \ldots, a_n\}$, of n integers for which no two are congruent modulo n is a complete residue system modulo n.*

Linear congruences

In the first chapter, we discussed some questions about finding solutions to linear Diophantine equations. Now we are going to take up analogous questions about finding solutions to equations in modular arithmetic. Specifically, our next goal is to determine when there are solutions to the general linear congruence

$$ax \equiv b \quad (\text{mod } n)$$

and how to find all the solutions. A solution is an integer value for x that makes the congruence true. We'll start with some examples.

3.18 Exercise. *Find all solutions in the appropriate canonical complete residue system modulo n that satisfy the following linear congruences:*

1. $26x \equiv 14 \pmod{3}$.

2. $2x \equiv 3 \pmod{5}$.

3. $4x \equiv 7 \pmod{8}$.

4. $24x \equiv 123 \pmod{213}$. *(This congruence is tedious to do by trial and error, so perhaps we should defer work on it for now and instead try to develop some techniques that might help.)*

This next theorem clearly connects the question of how to solve linear congruences with the techniques of solving linear Diophantine equations that we developed in Chapter 1.

3.19 Theorem. *Let a, b, and n be integers with $n > 0$. Show that $ax \equiv b$ (mod n) has a solution if and only if there exist integers x and y such that $ax + ny = b$.*

These theorems will encourage you to remember your work from Chapter 1.

3.20 Theorem. *Let* a, b, *and* n *be integers with* $n > 0$. *The equation* $ax \equiv b \pmod{n}$ *has a solution if and only if* $(a, n) | b$.

Now we have a specific condition that tells whether a linear congruence will or will not have a solution. We can use this criterion to see whether our deferred congruence in Exercise 3.18 does or does not have a solution.

3.21 Question. *What does the preceding theorem tell us about the congruence* (4) *in Exercise* 3.18 *above?*

Now let's actually solve the congruence in a systematic way. As usual, this work is tying back into the work we did in solving linear Diophantine equations in Chapter 1.

3.22 Exercise. *Use the Euclidean Algorithm to find a member* x *of the canonical complete residue system modulo* 213 *that satisfies* $24x \equiv 123$ (mod 213). *Find all members* x *of the canonical complete residue system modulo* 213 *that satisfy* $24x \equiv 123$ (mod 213).

Having done a specific example, as usual we step back and try to describe a general procedure.

3.23 Question. *Let* a, b, *and* n *be integers with* $n > 0$. *How many solutions are there to the linear congruence* $ax \equiv b$ (mod n) *in the canonical complete residue system modulo* n? *Can you describe a technique to find them?*

The next theorem gives the answer, so try to think it through on your own before reading on. While thinking about this question, crystallizing the ideas about linear Diophantine equations will help.

3.24 Theorem. *Let* a, b, *and* n *be integers with* $n > 0$. *Then*

1. *The congruence* $ax \equiv b$ (mod n) *is solvable in integers if and only if* $(a, n) | b$;

2. *If* x_0 *is a solution to the congruence* $ax \equiv b$ (mod n), *then all solutions are given by*

$$x_0 + \left(\frac{n}{(a, n)} \cdot m \right) \pmod{n}$$

for $m = 0, 1, 2, \ldots, (a, n) - 1$; *and*

3. *If* $ax \equiv b$ (mod n) *has a solution, then there are exactly* (a, n) *solutions in the canonical complete residue system modulo* n.

Systems of linear congruences: the Chinese Remainder Theorem

Sometimes in real life, we are confronted with problems involving simultaneous linear congruences. Something like the following has probably happened to you.

3.25 Exercise. *A band of 17 pirates stole a sack of gold coins. When they tried to divide the fortune into equal portions, 3 coins remained. In the ensuing brawl over who should get the extra coins, one pirate was killed. The coins were redistributed, but this time an equal division left 10 coins. Again they fought about who should get the remaining coins and another pirate was killed. Now, fortunately, the coins could be divided evenly among the surviving 15 pirates. What was the fewest number of coins that could have been in the sack?*

Perhaps your experience is less violent and more bucolic. Eggs need counting too.

3.26 Exercise (Brahmagupta, 7th century A.D.). *When eggs in a basket are removed two, three, four, five or six at a time, there remain, respectively, one, two, three, four, or five eggs. When they are taken out seven at a time, none are left over. Find the smallest number of eggs that could have been contained in the basket.*

These exercises are challenging but fun to do. The question now is whether we can formulate general statements that tell us when solutions to such problems exist and how those solutions can be found. This first theorem gives a criterion for when we can find a single number that is congruent to two different values modulo two different moduli. That single number is called a solution to a system of two linear congruences. Later we will consider solutions to arbitrarily large systems of linear congruences.

3.27 Theorem. *Let a, b, m, and n be integers with $m > 0$ and $n > 0$. Then the system*

$$x \equiv a \pmod{n}$$
$$x \equiv b \pmod{m}$$

has a solution if and only if $(n, m)|a - b$.

The next theorem asserts that in the case where $(m, n) = 1$, the solution is unique modulo the product mn.

3.28 Theorem. *Let a, b, m, and n be integers with $m > 0$, $n > 0$, and $(m, n) = 1$. Then the system*

$$x \equiv a \pmod{n}$$
$$x \equiv b \pmod{m}$$

has a unique solution modulo mn.

The most famous theorem along these lines is the Chinese Remainder Theorem. Here the moduli are relatively prime, but there can be any finite number of them. The pirate problem is a Chinese Remainder Theorem problem in disguise (possibly with an eye patch). The Chinese Remainder Theorem involves L different linear congruences. Whenever you see a theorem or a problem that has a potentially large natural number involved, it is a good idea to start thinking about the cases where L is 1 or 2 or 3. Doing those special cases is a great way to teach yourself how to do the general case. The previous theorem gets you started by doing the case $L = 2$. Also, you might think about induction in trying to then do the general case.

3.29 Theorem (Chinese Remainder Theorem). *Suppose $n_1, n_2, \ldots, n_L$ are positive integers that are pairwise relatively prime, that is, $(n_i, n_j) = 1$ for $i \neq j$, $1 \leq i, j \leq L$. Then the system of L congruences*

$$x \equiv a_1 \pmod{n_1}$$
$$x \equiv a_2 \pmod{n_2}$$
$$\vdots$$
$$x \equiv a_L \pmod{n_L}$$

has a unique solution modulo the product $n_1 n_2 n_3 \cdots n_L$.

3.30 Blank Paper Exercise. *After not looking at the material in this chapter for a day or two, take a blank piece of paper and outline the development of that material in as much detail as you can without referring to the text or to notes. Places where you get stuck or can't remember highlight areas that may call for further study.*

A Prince and a Master

Carl Friedrich Gauss, sometimes called the Prince of Mathematics, is considered by many to be one of the greatest mathematicians in history, and it is to him that we owe the modern theory and notation of congruences (i.e., modular arithmetic). His treatise *Disquisitiones Arithmeticae*, published in 1801 when Gauss was just 24, brought together for the first time in one source the important number theory contributions of many previous mathematicians, including Fermat, Euler, Joseph Lagrange, and Adrien-Marie Legendre. Some of Gauss' own contributions to number theory will be treated in later chapters.

Sun Zi wrote the Chinese treatise *Sun Tze Suan Ching*, which translates to *Master Sun's Mathematical Manual*. He is assumed to have lived during either the third or fourth century AD. There is some evidence that he was a Buddhist monk, but little else is known of him. Master Sun's manual is divided into three volumes, and Problem 26 from Volume 3 is translated

> We have a number of things, but we do not know exactly how many. If we count them by threes we have two left over. If we count them by fives we have three left over. If we count them by sevens we have two left over. How many things are there?

You will of course recognize this as a problem requiring a solution to a system of linear congruences, not unlike Brahmagupta's egg basket problem. It is because Sun Zi's text provides the earliest known example of such a problem that the Chinese Remainder Theorem obtained its name.

4

Fermat's Little Theorem and Euler's Theorem

Abstracting the Ordinary

One way that mathematics is created is to abstract, change, or generalize some features of familiar mathematical objects and see what happens. For example, we started with the familiar idea of arithmetic with integers and then made some changes to consider modular arithmetic, a sort of cyclical version of arithmetic. Abstract algebra is a mathematical exploration of generalizations of various familiar ideas such as the integers, the rational numbers, the real numbers and their associated arithmetic operations and properties. By selectively focusing on some properties of these examples, abstract algebra constructs categories of algebraic entities including objects called groups, rings, and fields. Modular arithmetic provides us with examples of some of these algebraic structures and illustrates some of the properties that lead to many fundamental ideas in abstract algebra.

Solving the linear congruence

$$ax \equiv b \pmod{n}$$

means finding a number that when added to itself a times results in b modulo n. In studying such congruences we are implicitly studying the results of repeated addition modulo n and patterns that this process might produce. Equally interesting, as well as fruitful, is the study of repeated multiplication modulo n, that is, taking powers of numbers and reducing those powers modulo n. The operations of addition and multiplication are so well understood in the natural numbers that looking at their behavior in modular arithmetic is a natural exploration to undertake.

Orders of an integer modulo n

We begin here by exploring how powers of numbers behave modulo n. We will find a structure among numbers modulo n that is interesting in its own right, has applications in cryptography and coding among other fields, and leads to central ideas of group theory. As usual we will do some specific examples in order to help us develop some intuition about what we might expect.

4.1 Exercise. *For $i = 0$, 1, 2, 3, 4, 5, and 6, find the number in the canonical complete residue system to which 2^i is congruent modulo 7. In other words, compute 2^0 (mod 7), 2^1 (mod 7), 2^2 (mod 7), ..., 2^6 (mod 7).*

Taking powers of an integer cannot create common factors with another integer if none existed to start with.

4.2 Theorem. *Let a and n be natural numbers with $(a, n) = 1$. Then $(a^j, n) = 1$ for any natural number j.*

Reducing a number modulo n cannot create a common factor with n.

4.3 Theorem. *Let a, b, and n be integers with $n > 0$ and $(a, n) = 1$. If $a \equiv b$ (mod n), then $(b, n) = 1$.*

If you raise a number to various powers, you will sometimes get the same values modulo n.

4.4 Theorem. *Let a and n be natural numbers. Then there exist natural numbers i and j, with $i \neq j$, such that $a^i \equiv a^j$ (mod n).*

The next theorem repeats a theorem we saw before, but it is one of the most used theorems in the exploration of powers, so you should have its statement and proof at the tips of your fingers.

4.5 Theorem. *Let a, b, c, and n be integers with $n > 0$. If $ac \equiv bc$ (mod n) and $(c, n) = 1$, then $a \equiv b$ (mod n).*

The next theorem tells us that if we take a natural number relatively prime to a modulus n, then some power of it will be congruent to 1 modulo n. One consequence of this theorem is that after a power gets to 1, the powers will just recycle.

4.6 Theorem. *Let a and n be natural numbers with $(a, n) = 1$. Then there exists a natural number k such that $a^k \equiv 1$ (mod n).*

The preceding theorem tells us that every natural number relatively prime to a modulus has an exponent naturally associated with it, namely, the smallest exponent that makes the power congruent to 1. That concept is so useful that we give it a name.

Definition. Let a and n be natural numbers with $(a, n) = 1$. The smallest natural number k such that $a^k \equiv 1 \pmod{n}$ is called the *order of a modulo n* and is denoted $\text{ord}_n(a)$.

Fermat's Little Theorem

The culminating theorem of this section is Fermat's Little Theorem. It gives us information about what power of a number will be congruent to 1 modulo a prime. We will approach that theorem by first finding some sort of a bound on the size of the order of a natural number. Experimenting with some actual numbers is a good way to begin.

4.7 Question. *Choose some relatively prime natural numbers a and n and compute the order of a modulo n. Frame a conjecture concerning how large the order of a modulo n can be, depending on n.*

In doing your experiments of taking a number to powers, you might have noticed that until the power was congruent to 1 modulo n, the values modulo n never repeated. That observation is the content of the next theorem.

4.8 Theorem. *Let a and n be natural numbers with $(a, n) = 1$ and let $k = \text{ord}_n(a)$. Then the numbers a^1, a^2, ..., a^k are pairwise incongruent modulo n.*

Taking powers of a natural number beyond its order will never produce different numbers modulo n.

4.9 Theorem. *Let a and n be natural numbers with $(a, n) = 1$ and let $k = \text{ord}_n(a)$. For any natural number m, a^m is congruent modulo n to one of the numbers a^1, a^2, ..., a^k.*

The only powers of a natural number that give 1 modulo n are powers that are multiples of the order.

4.10 Theorem. *Let a and n be natural numbers with $(a, n) = 1$, let $k = \text{ord}_n(a)$, and let m be a natural number. Then $a^m \equiv 1 \pmod{n}$ if and only if $k \mid m$.*

This next theorem may have been what you conjectured when you did your experiments about order in the first question of this section. It states that the order of a natural number, that is, the power that first gets you to 1 modulo n, is less than n.

4.11 Theorem. *Let a and n be natural numbers with $(a, n) = 1$. Then* $\text{ord}_n(a) < n$.

The following question asks you to do some experiments that might lead you to make a conjecture about powers of numbers modulo primes. You will probably make the conjecture that we will see later is in fact a theorem, Fermat's Little Theorem.

4.12 Exercise. *Compute a^{p-1} (mod p) for various numbers a and primes p, and make a conjecture.*

The numbers $1, 2, 3, \ldots, p$ form a complete residue system modulo p. The next theorem states that if p is a prime, then multiplying each of those numbers by a fixed number that is not divisible by p produces another complete residue system. You might want to take a small prime, like 5, and multiply each of the numbers $1, 2, 3, 4, 5$ by some other number, for example, 6, and check that you produce a complete residue system.

4.13 Theorem. *Let p be a prime and let a be an integer not divisible by p; that is, $(a, p) = 1$. Then $\{a, 2a, 3a, \ldots, pa\}$ is a complete residue system modulo p.*

Multiplying all the natural numbers less than a prime p will give the same result modulo p as multiplying a fixed multiple of those numbers.

4.14 Theorem. *Let p be a prime and let a be an integer not divisible by p. Then*

$$a \cdot 2a \cdot 3a \cdots (p-1)a \equiv 1 \cdot 2 \cdot 3 \cdots (p-1) \pmod{p}.$$

This theorem can be used to prove Fermat's Little Theorem, which follows. We state two versions of Fermat's Little Theorem, but ask you to prove that the two versions are equivalent to one another. Both of them tell us important and applicable facts about powers of natural numbers modulo a prime.

4.15 Theorem (Fermat's Little Theorem, Version I). *If p is a prime and a is an integer relatively prime to p, then $a^{(p-1)} \equiv 1 \pmod{p}$.*

4.16 Theorem (Fermat's Little Theorem, Version II). *If p is a prime and a is any integer, then $a^p \equiv a$ (mod p).*

4.17 Theorem. *The two versions of Fermat's Little Theorem stated above are equivalent to one another, that is, each one can be deduced from the other.*

Fermat's Little Theorem states that a natural number not divisible by p, raised to the $(p-1)$-st power, is congruent to 1 modulo p. Recall that the order of a natural number is the smallest power that is congruent to 1 modulo p. The next theorem states that the order of each such number must divide $(p-1)$.

4.18 Theorem. *Let p be a prime and a be an integer. If $(a, p) = 1$, then $\mathrm{ord}_p(a)$ divides $p-1$, that is, $\mathrm{ord}_p(a) | p - 1$.*

One of the impressive applications of Fermat's Little Theorem is that it allows us to do computations involving modular arithmetic that would be impossible otherwise. Impress your friends by doing the following computations in your head.

4.19 Exercise. *Compute each of the following without the aid of a calculator or computer.*
   1. 512^{372} (mod 13).
   2. 3444^{3233} (mod 17).
   3. 123^{456} (mod 23).

4.20 Exercise. *Find the remainder upon division of 314^{159} by 31.*

Fermat's Little Theorem tells us information about prime moduli, but how are we going to deal with moduli that are not prime? One strategy is to decompose a composite (non-prime) modulus into relatively prime parts. The following theorem shows that a natural number that is congruent to a fixed number modulo two different, relatively prime moduli is congruent to that same number modulo the product of the moduli. For example, if you have a natural number that is congruent to 12 modulo 15 and that same number is congruent to 12 modulo 8, that number is also congruent to 12 modulo 120 (= 8 · 15).

4.21 Theorem. *Let n and m be natural numbers that are relatively prime, and let a be an integer. If $x \equiv a$ (mod n) and $x \equiv a$ (mod m), then $x \equiv a$ (mod nm).*

4.22 Exercise. *Find the remainder when* 4^{72} *is divided by* $91\ (=7\cdot 13)$.

When you see powers and a modulus, it is a good idea to think about the modulus as a product of primes and then see whether you can use Fermat's Little Theorem to advantage.

4.23 Exercise. *Find the natural number* $k < 117$ *such that* $2^{117} \equiv k$ (mod 117). *(Notice that* 117 *is not prime.)*

An alternative route to Fermat's Little Theorem

Many theorems have several different proofs. One approach to proving Fermat's Little Theorem is by induction using the Binomial Theorem. So the first step in this approach is to state and prove the Binomial Theorem.

Definition. If n and m are natural numbers with $m \leq n$, then

$$\binom{n}{m} = \frac{n!}{m!(n-m)!}.$$

We define 0! to equal 1. Thus, we can extend the definition to include $m = 0$. In that case, we have $\binom{n}{0} = 1$ for any natural number n.

Note: You may recall that $\binom{n}{m}$ is equal to the number of subsets of size m in a set of size n.

4.24 Theorem (Binomial Theorem). *Let a and b be numbers and let n be a natural number. Then*

$$(a+b)^n = \sum_{i=0}^{n} \binom{n}{i} a^{n-i} b^i.$$

The Binomial Theorem describes the coefficients of each term when you expand $(a+b)^n$. When n is equal to a prime p, p will divide all those coefficients, except the end ones, of course.

4.25 Lemma. *If p is prime and i is a natural number less than p, then p divides $\binom{p}{i}$.*

Using this observation, you can prove Fermat's Little Theorem, Version II, by first observing that 0^p is congruent to 0 modulo p, 1^p is congruent to 1 modulo p, then moving on to prove that 2^p is congruent to 2 modulo p and then proving that 3^p is congruent to 3 modulo p and so on. You might find the preceding lemma useful in executing this inductive procedure.

4.26 Theorem (Fermat's Little Theorem, Version II). *If p is a prime and a is an integer, then $a^p \equiv a$ (mod p).*

Euler's Theorem and Wilson's Theorem

Fermat's Little Theorem suffers from the limitation that the modulus is prime. As usual, our strategy is to take an idea, in this case Fermat's Little Theorem, and see how it can be extended to apply to a more general case. So we need to ask ourselves what aspects of Fermat's Little Theorem can we hope to extend to a case where the modulus is not prime. If we start with a number that is not relatively prime to the modulus, then no power of it will ever be congruent to 1. So we focus our attention on those numbers that are relatively prime to the modulus. The first concept we introduce is the Euler ϕ-function that simply counts how many of these relatively prime numbers there are.

Definition. For a natural number n, the *Euler ϕ-function*, $\phi(n)$, is equal to the number of natural numbers less than or equal to n that are relatively prime to n. (Note that $\phi(1) = 1$.)

Let's just do a few examples to make sure that the definition is clear.

4.27 Question. *The numbers 1, 5, 7, and 11 are all the natural numbers less than or equal to 12 that are relatively prime to 12, so $\phi(12) = 4$.*

  1. What is $\phi(7)$?

  2. What is $\phi(15)$?

  3. What is $\phi(21)$?

  4. What is $\phi(35)$?

It is always a good idea to revisit useful and important results and remind yourself of their proofs. We restate the following three theorems here because of their importance and usefulness in the upcoming work.

4.28 Theorem. *Let a, b, and n be integers such that $(a,n) = 1$ and $(b,n) = 1$. Then $(ab,n) = 1$.*

4.29 Theorem. *Let a, b, and n be integers with $n > 0$. If $a \equiv b$ (mod n) and $(a,n) = 1$, then $(b,n) = 1$.*

4.30 Theorem. *Let a, b, c, and n be integers with n > 0. If ab ≡ ac* (mod *n*) *and* $(a, n) = 1$, *then* $b \equiv c$ (mod *n*).

The following theorem begins by listing those numbers that are being counted when we find the Euler ϕ-function of a number. It observes that multiplying each of those numbers by a common number that is relatively prime to the modulus cannot create congruent numbers. They start not congruent (because they are different numbers less than the modulus) and they end not congruent.

4.31 Theorem. *Let n be a natural number and let* $x_1, x_2, \ldots, x_{\phi(n)}$ *be the distinct natural numbers less than or equal to n that are relatively prime to n. Let a be a non-zero integer relatively prime to n and let i and j be different natural numbers less than or equal to* $\phi(n)$. *Then* $ax_i \not\equiv ax_j$ (mod *n*).

The next theorem is Euler's Theorem, which generalizes Fermat's Little Theorem. Since Euler's Theorem generalizes Fermat's Little Theorem, the way to start thinking about its proof is to think about the proof of Fermat's Little Theorem and see whether you can imitate the steps in this different setting. It is always a good idea to start with what you know and see how it can be modified to fit a new situation.

4.32 Theorem (Euler's Theorem). *If a and n are integers with n > 0 and* $(a, n) = 1$, *then*

$$a^{\phi(n)} \equiv 1 \pmod{n}.$$

4.33 Corollary (Fermat's Little Theorem). *If p is a prime and a is an integer relatively prime to p, then* $a^{(p-1)} \equiv 1$ (mod *p*).

As long as we can compute $\phi(n)$, Euler's Theorem can be used just like Fermat's Little Theorem for computing powers of numbers modulo *n*.

4.34 Exercise. *Compute each of the following without the aid of a calculator or computer.*

1. 12^{49} (mod 15).

2. 139^{112} (mod 27).

4.35 Exercise. *Find the last digit in the base* 10 *representation of the integer* 13^{474}.

The next theorem tells us that every natural number less than a given prime can be multiplied by another natural number to yield 1 modulo the prime. This observation says that numbers have something that behaves like a multiplicative inverse in the "mod p" world.

4.36 Theorem. *Let p be a prime and let a be an integer such that $1 \leq a < p$. Then there exists a unique natural number b less than p such that $ab \equiv 1 \pmod{p}$.*

Definition. Let p be a prime and let a and b be integers such that $ab \equiv 1 \pmod{p}$. Then a and b are said to be *inverses modulo p.*

4.37 Exercise. *Let p be a prime. Show that the natural numbers 1 and $p - 1$ are their own inverses modulo p.*

The next theorem asserts that except for the special numbers 1 and $p-1$, the inverse of a number modulo p is different from itself. In other words, squaring a natural number less than p other than 1 or $p - 1$ will not give you a number congruent to 1 modulo the prime p.

4.38 Theorem. *Let p be a prime and let a and b be integers such that $1 < a, b < p - 1$ and $ab \equiv 1 \pmod{p}$. Then $a \neq b$.*

Let's see how numbers pair up with their inverses in a specific case.

4.39 Exercise. *Find all pairs of numbers a and b in $\{2, 3, \ldots, 11\}$ such that $ab \equiv 1 \pmod{13}$.*

The preceding theorems and examples are giving us a perspective about numbers and their multiplicative inverses modulo a prime p. One consequence of this picture is that when we multiply all the numbers from 2 up to $(p - 2)$, we get a number congruent to 1 modulo the prime p.

4.40 Theorem. *If p is a prime larger than 2, then $2 \cdot 3 \cdot 4 \cdots \cdot (p - 2) \equiv 1 \pmod{p}$.*

We end the chapter with Wilson's Theorem which is perhaps the most famous consequence of our understanding of numbers and their inverses modulo a prime p.

4.41 Theorem (Wilson's Theorem). *If p is a prime, then $(p - 1)! \equiv -1 \pmod{p}$.*

The converse of Wilson's Theorem is also true; that is, if the product of all the natural numbers less than n is congruent to -1 modulo n, then n must be prime.

4.42 Theorem (Converse of Wilson's Theorem). *If n is a natural number such that*

$$(n-1)! \equiv -1 \pmod{n},$$

then n is prime.

Whenever we prove a good theorem, we can ask about extensions of it. After we proved Fermat's Little Theorem that talked about prime moduli, we extended it to Euler's Theorem that dealt with composite moduli. Can you make a conjecture that would extend Wilson's Theorem to moduli that are not prime?

4.43 Blank Paper Exercise. *Chapter 4 is the culmination of all of your inquiries from the first three chapters. After not looking at the material for a day or two, take a blank piece of paper and outline the development of the first four chapters in as much detail as you can without referring to the text or to notes. Places where you get stuck or can't remember highlight areas that may call for further study.*

Fermat, Wilson and ... Leibniz?

Tracing the history of named results like those of this chapter can be trying. Shakespeare's famous "What's in a name?" aptly applies. In a letter to Frenicle de Bessy (1605–1675) dated 1640, Fermat stated what we now call Fermat's Little Theorem. Characteristic of Fermat, the theorem was explained without proof stating "I would send you the demonstration, if I did not fear its being too long."

It is not until 1736 that we find the first published proof in the works of Euler. The argument is based on the Binomial Theorem, and could likely have been known to Fermat. The algebraic proof given in Theorems 4.13–4.15 appeared in 1806, and is attributed to James Ivory (1765–1842). Euler, of course, went on to generalize Fermat's Little Theorem and published a proof of Euler's Theorem in 1760.

Abu Ali al-Hasan ibn al-Haytham (approx. 965–1040) considered the following problem: *To find a number such that if we divide by two, one remains; if we divide by three, one remains; if we divide by four, one remains; if we divide by five, one remains; if we divide by six, one remains; if we divide by seven, there is no remainder.* His method of solution gives, in this particular case, the number $(7 - 1)! + 1$, which clearly leaves a remainder of 1 upon division by 2, 3, 4, 5 and 6. But al-Haytham was also aware that this number was divisible by 7, which is an instance of Wilson's theorem.

Nearly 800 years later Edward Waring (1736–1798) first published the general statement of Wilson's Theorem, attributing the result to his student John Wilson (1741–1793). No proof was given in Waring's publication, and it is believed that neither Waring nor Wilson were aware of a proof. The first published proof, based on the binomial theorem, appeared in 1773 by Lagrange and also included a proof of the converse of Wilson's Theorem.

Enter Leibniz. In 1894 attention was called to a collection of unpublished manuscripts located in the Hanover Library attributed to Gottfried Wilhelm von Leibniz (1646–1716), most famous as one of the creators of Calculus as well as for his philosophical theory of monads. We usually do not think of Leibniz as a pioneer of number theory. However, among his works found in the Hanover Library are results believed to have been attained prior to 1683 which include proofs of both Fermat's Little Theorem and Wilson's Theorem. These dates precede Euler's first published proof of Fermat's Little Theorem by 53 years and Lagrange's first published proof of Wilson's Theorem by 90 years.

5

Public Key Cryptography

Public Key Codes and RSA

Public key codes

Public key codes are codes in which the encoding method is public knowledge; i.e., anyone can encode messages. However, even though everybody knows how messages are encoded, only the receiver knows how to decode an encrypted message. For example, suppose I want to sell a product and I want customers to be able to send me their credit card numbers in a secure manner. I can "publish" a public encoding scheme. People use this scheme to encode their credit card numbers before sending them to me. For the scheme to be secure, I should be the only person who can decode the numbers. So even though everyone knows *exactly* how the numbers were encoded, only I can "undo" the encoding in order to decode the message.

Such codes are called *public key codes*. The notion is counterintuitive. How can such a scheme work? The answer is based on the fact that certain mathematical operations are easy to perform, but hard to undo. We will look at a specific public key encoding scheme called RSA encryption, first created by mathematicians Ronald Rivest, Adi Shamir, and Leonard Adleman.

Overview of RSA

Suppose we select two enormous prime numbers, each on the order of 200 digits long, for example. Now we multiply them (computers are whizzes at multiplying natural numbers, even numbers with hundreds of digits). Now we give our result to a friend and ask her to factor it. She goes off to have her computer help her out, and is never seen again. *Factoring* large numbers is hard, even for a computer. There are limits to the size of natural numbers that a computer can factor. Our product of two 200 digit primes is much too large for even the fastest computers to factor.

So we can announce our enormous number to the world, but only we know its factors. At this point, you would be justified in saying, "So what? Who cares what the factors of a 400 digit number are anyway?" The answer is that you care. You care because the inability to factor such numbers is at the heart of public key encryption systems that are used millions of times a day to keep data that is sent over the internet secure. The challenge for this chapter is for you to discover how to make a public key code system by exploiting this example of a mathematical operation that is easy to perform (the multiplication of two large primes), but hard to undo (factor). We will see how the huge product is the *public* part of the RSA encryption scheme that will somehow allow anyone to encode messages while the decoding requires knowing its factorization, thus making the code unbreakable except by the person who knows the factors. Of course, at this point there is no apparent connection between factoring numbers and encoding messages. That is the content of this chapter.

For convenience, let's suppose the message we wish to encode is a number. If our message contained words, we could do some sort of simple transformation turning letters into numbers. We will take our message number and perform a mathematical operation on it to produce a new number. This new number is the encoded message. What operation will we perform? We will raise our original number message to some power modulo some base. Recovering the original number message from the encoded message number will be practically impossible without some secret knowledge. With the secret knowledge, we simply raise the encoded number to another power to obtain the original message. The key to the whole process is the work we have already done, including the Euclidean Algorithm and Euler's Theorem.

Let's decrypt

Before getting to James Bond, let's begin with some theorems about modular arithmetic. This first theorem has a familiar conclusion reminiscent of Fermat's Little Theorem and Euler's Theorem, namely, that under certain conditions a number to a power is congruent to 1 modulo another number.

5.1 Theorem. *If p and q are distinct prime numbers and W is a natural number with $(W, pq) = 1$, then $W^{(p-1)(q-1)} \equiv 1 \pmod{pq}$.*

You might think that the next theorem would require the hypothesis that $(W, pq) = 1$; however, it is true for all natural numbers W. One strategy for proving a theorem is first to prove the theorem with a stronger hypothesis and later deal with the other cases. Here, you might first prove the theorem assuming the extra hypothesis that $(W, pq) = 1$. After that success, you can analyze what would happen if p or q divides W.

5.2 Theorem. *Let p and q be distinct primes, k be a natural number, and W be a natural number less than pq. Then*

$$W^{1+k(p-1)(q-1)} \equiv W \pmod{pq}.$$

Notice how this next theorem has a conclusion that looks similar to theorems from Chapter 1 about linear Diophantine equations. As usual, an excellent strategy in mathematics is to remember previous theorems or insights that seem to be related to the current question.

5.3 Theorem. *Let p and q be distinct primes and E be a natural number relatively prime to $(p-1)(q-1)$. Then there exist natural numbers D and y such that*

$$ED = 1 + y(p-1)(q-1).$$

5.4 Theorem. *Let p and q be distinct primes, W be a natural number less than pq, and E, D, and y be natural numbers such that $ED = 1 + y(p-1)(q-1)$. Then*

$$W^{ED} \equiv W \pmod{pq}.$$

Notice that the conclusion of the preceding theorem is that raising W to a certain power, the ED power, and reducing modulo pq just gives us W back again. Remember that $W^{ED} = (W^E)^D$.

We now have all the pieces used to make up the RSA Public Key Coding System. The next exercise asks you to put the pieces together.

5.5 Exercise. *Consider two distinct primes p and q. Describe every step of the RSA Public Key Coding System. State what numbers you choose to make public, what messages can be encoded, how messages should be encoded, and how messages are decoded. What number should be called the encoding exponent and what number should be called the decoding exponent?*

The next exercise asks you to develop an RSA Public Key Coding System using an actual pair of primes. These primes might be slightly too small for any real value in applications, but the goal of the exercise is for you to understand every step of how the RSA system works and see it actually work with numbers. Again, state what numbers you choose to make public, what messages can be encoded, how messages should be encoded, and how messages are decoded. It is neat to see all these steps and to see that you can encode and decode actual numbers.

5.6 Exercise. *Describe an RSA Public Key Code System based on the primes* 11 *and* 17. *Encode and decode several messages.*

Of course, the fun of being a spy is to break codes. So get on your trench coat, pull out your magnifying glass, and begin to spy. The next exercise asks you to break an RSA code and save the world.

5.7 Exercise. *You are a secret agent. An evil spy with shallow number theory skills uses the RSA Public Key Coding System in which the public modulus is* $n = 1537$, *and the encoding exponent is* $E = 47$. *You intercept one of the encoded secret messages being sent to the evil spy, namely the number* 570. *Using your superior number theory skills, decode this message, thereby saving countless people from the fiendish plot of the evil spy.*

The next exercise asks you to explain in general how you can break RSA codes if you are able to factor n.

5.8 Exercise. *Suppose an RSA Public Key Coding System publishes* n *(which is equal to the product of two undisclosed primes* p *and* q) *and* E, *with* E *relatively prime to* $(p-1)(q-1)$. *Suppose someone wants to send a secret message and so encodes the message number* W *(less than* n) *by finding the number* m *less than* n *such that* $m \equiv W^E$ *(mod* n). *Suppose you intercept this number* m *and you are able to factor* n. *How can you figure out the original message* W?

Notice that the two previous exercises tell us that the RSA Public Key Coding System would be useless if it were possible to factor pq. Factoring sounds like a simple process; however, when p and q are primes containing several hundred digits each, no person nor computer in the world knows how to factor pq. It is interesting that such a simple process as factoring lies at the heart of secret codes on which billions of dollars of secure transactions rely.

5.9 Applications Exercise. *You have seen the application of number theory to RSA cryptography. Find out all you can about the role of number theory in some other types of "codes" such as bar codes, ISBN codes, and credit card number "codes."*

Hard Problems

The RSA encryption system actually has two keys. One is made public (the encoding key E), and the other is kept private (the decoding key D). Such a system is said to use an *asymmetrical* key, as opposed to a *symmetrical* key where the same key is used to both encrypt and decrypt. The asymmetrical public key allows anyone to encode messages, but only the receiver can decode. In practice, the RSA system is inefficient for encoding and decoding large amounts of data. Encryption methods such as AES (Advanced Encryption Standard) are much more efficient, but require a symmetric key to be shared by the sender and receiver. Sharing such a key poses many potential problems. So we have

- AES: efficient, but requires a shared key,

- RSA: inefficient, but uses a public key.

In practice, the two methods are often combined to take advantage of their positive qualities (the efficiency of AES and the public key of RSA).

If Alice wishes to send a message M to Bob, she encrypts M using a randomly chosen AES key. Then, using Bob's public RSA encoding key, she encrypts her AES key. Alice then sends Bob two items: her AES encoded message and her RSA encrypted AES key. Bob can easily decrypt the AES key (using his private RSA decryption key), then use the decrypted AES key to decrypt the AES encoded message. So in this regard, RSA is primarily used as a method of *key exchange*.

The security of the RSA encryption system relies on the fact that factoring is hard. How hard? According to the RSA Laboratories website, it was reported in November of 2005 that a 193 digit integer was factored after 30 2.2GHz-Opteron-CPU years of work (which occurred over about 5 months of calendar time). We're not exactly sure what that statement means, but it sure makes factoring sound hard. But factoring is not the only hard mathematical problem used for *public key exchange*.

Some of the earliest work on public key exchange methods occurred in the mid-1970s at Stanford University. Graduate student Whitfield Diffie and his advisor Martin Hellman developed a public key exchange system based on the hard mathematical problem of computing "logarithms modulo p." It works as follows. Suppose Alice and Bob wish to share a secret key (which will simply be a number). Two quantities are made public: a prime number p, and an integer $g < p$ which has the property that $\{0, g, g^2, \ldots, g^{p-1}\}$ form a complete residue system modulo p. Such a g is called a *primitive root modulo p*, and is explored further in the next chapter.

Next, Alice and Bob each choose a private value, say a and b. These numbers are not made public. Alice then makes public her value g^a (mod p), and Bob makes public his value g^b (mod p). Finally, Alice and Bob can now compute their shared secret key: Alice takes Bob's public value and computes $(g^b)^a$ (mod p), and Bob takes Alice's public value and computes $(g^a)^b$ (mod p). Since

$$(g^b)^a \equiv g^{ba} \equiv g^{ab} \equiv (g^a)^b \pmod{p},$$

they have a shared key (which, for example, could then be used for a symmetrical key system like AES). How secret is it? Essentially, the only way to figure out the shared key is to obtain the secret values a and b. So the problem becomes: given the public values g and g^a (mod p), determine the secret value a. This is called the *discrete logarithm problem modulo p*, and it is believed to be just as difficult as the factoring problem associated with RSA.

The group of integers modulo n are not the only source of mathematics making its way into public key cryptography. In the mid-1980s Victor Miller and Neal Koblitz independently proposed using mathematical objects called *elliptic curves* to generate public key codes. An elliptic curve is a plane cubic curve. For example, an elliptic curve might be given by an equation of the form

$$y^2 = x^3 + bx + c,$$

where b and c are chosen from an appropriate set of numbers. What is special about these curves is that they come with an arithmetic as well. That is, there is a natural way to "add" two points on the curve and obtain a third point.

As with Diffie-Hellman, certain objects are made public: the elliptic curve, a prime number p, and a "point" P on the elliptic curve. The prime p specifies where the coefficients b and c in the equation of our curve are coming from. Namely, they come from the set of integers modulo p, i.e., the set $\{0, 1, 2, \ldots, p-1\}$. The point P is then an ordered pair $P = (x, y)$ where x and y are integers modulo p that satisfy the curve's equation modulo p; that is, x and y satisfy

$$y^2 \equiv x^3 + bx + c \pmod{p}.$$

For example, consider the following curve with coefficients coming from the set of integers modulo 23 (so $p = 23$): $y^2 = x^3 + x$. It is a good exercise to check that $P = (17, 13)$ is in fact a "point" on the curve (there are actually 23 "points" on this curve modulo 23).

Alice uses her secret value a to compute a public "point"

$$aP = \underbrace{P + P + \cdots + P}_{a \text{ terms}},$$

and Bob makes public bP. They can then compute their shared secret key

$$a(bP) = (ab)P = (ba)P = b(aP).$$

For a third party to discover their secret key, the values a and b must be found. So the problem becomes: given the public quantities P and aP, find a. This is the *discrete logarithm problem for elliptic curves modulo p*, and is currently considered a harder problem than the discrete logarithm problem for the integers modulo p that provides the security for Diffie-Hellman.

These public key coding systems use abstract results in number theory to do the very practical work of sending messages over the internet. When mathematicians were working on the underlying number theory, they had no notion that their work would have any practical applications. Fermat and Euler, whose theorems are crucial to the public key coding messages we developed in this chapter, lived hundreds of years ago. They found the number theory results beautiful and interesting. Often mathematics has been developed without applications in mind and then later those insights are discovered to be crucial to some very important practical issue. Public key cryptography is a prime example of how important it is for human beings to continue to explore ideas in mathematics and science with the only goal being to seek and develop the beauty of ideas. Practical applications will inevitably follow.

6

Polynomial Congruences and Primitive Roots

Higher Order Congruences

The RSA coding system embodies a beautiful application of Euler's Theorem. A key step in the decoding process was our ability to solve the congruence $x^E \equiv m \pmod{pq}$, where E was the encoding exponent and m was the encoded word. This may have been our first example of a polynomial congruence of degree greater than 1 (recall we covered linear congruences back in Chapter 3). In this chapter and the next we continue the study of solutions to polynomial congruences of higher degree, encountering some fascinating new mathematics along the way.

Lagrange's Theorem

One of the most basic theorems about polynomials is the Fundamental Theorem of Algebra. Among other things, it tells us that an nth degree polynomial

$$f(x) = a_n x^n + a_{n-1} x^{n-1} + \cdots + a_0$$

has no more than n roots. We will not attempt to give a proof here of the Fundamental Theorem of Algebra. Rather, we will derive a "mod p" version of it due to Lagrange.

Definition. Recall that r is a *root* of the polynomial $f(x) = a_n x^n + a_{n-1} x^{n-1} + \cdots + a_0$ if and only if $f(r) = 0$.

This first theorem does not have any modular arithmetic in it. Do you remember how to do long division with polynomials?

6.1 Theorem. *Let $a_n x^n + a_{n-1} x^{n-1} + \cdots + a_0$ be a polynomial of degree $n > 0$ with integer coefficients and assume $a_n \neq 0$. Then an integer r is a root of $f(x)$ if and only if there exists a polynomial $g(x)$ of degree $n - 1$ with integer coefficients such that $f(x) = (x - r)g(x)$.*

This next theorem is very similar to the one above, but in this case $(x - r)g(x)$ is not quite equal to $f(x)$, but is the same except for the constant term of $f(x)$ and the constant term of $(x - r)g(x)$. Those constant terms are not the same, but are congruent using an appropriate modulus.

6.2 Theorem. *Let $f(x) = a_n x^n + a_{n-1} x^{n-1} + \cdots + a_0$ be a polynomial of degree $n > 0$ with integer coefficients and $a_n \neq 0$. Let p be a prime number and r an integer. Then, if $f(r) \equiv 0 \pmod{p}$, there exists a polynomial $g(x)$ of degree $n - 1$ such that*

$$(x - r)g(x) = a_n x^n + a_{n-1} x^{n-1} + \cdots + a_1 x + b_0$$

where $a_0 \equiv b_0 \pmod{p}$.

The final theorem of this section is a generalization of the Fundamental Theorem of Algebra in the setting of polynomials modulo a prime.

6.3 Theorem (Lagrange's Theorem). *If p is a prime and $f(x) = a_n x^n + a_{n-1} x^{n-1} + \cdots + a_0$ is a polynomial with integer coefficients and $a_n \neq 0$, then $f(x) \equiv 0 \pmod{p}$ has at most n non-congruent solutions modulo p.*

Primitive roots

Fermat's Little Theorem tells us that if we raise a natural number a less than a prime p to the $p - 1$ power, the result is congruent to 1 modulo p. However, for some natural numbers a, raising a to lower powers may also result in a number congruent to 1 modulo p. In this section, you will explore the orders of elements in more detail. Let's begin by proving that the order of a is the same as the order of a^i if i is relatively prime to the order.

6.4 Theorem. *Suppose p is a prime and $\mathrm{ord}_p(a) = d$. Then for each natural number i with $(i, d) = 1$, $\mathrm{ord}_p(a^i) = d$.*

The preceding theorem gives us a whole collection of numbers that have the same order modulo p. The next theorem, by contrast, puts a limit on how many incongruent natural numbers can have the same order modulo p. You might notice that a natural number k of order d modulo a prime

p is a solution of the congruence $x^d \equiv 1 \pmod{p}$. Recall that we earlier proved some theorems concerning the number of incongruent solutions that an equation of degree d could have modulo p.

6.5 Theorem. *For a prime p and natural number d, at most $\phi(d)$ incongruent integers modulo p have order d modulo p.*

Of course, there are many natural numbers d in the above theorem for which there are no numbers with that order modulo p. Recall that the order of any integer modulo p is less than p. In fact, recall the theorem that if p is a prime and k is a natural number less than p, then $\operatorname{ord}_p(k)|(p-1)$. It is always a good idea to review the proof or the main steps of the proof when you recall a theorem. In this case, you may remember something like the following key ideas. By definition of order, $k^{\operatorname{ord}_p(k)} \equiv 1 \pmod{p}$ and no lower power of k is congruent to 1 modulo p. Therefore, $k^{2\operatorname{ord}_p(k)} \equiv 1 \pmod{p}$ and $k^{3\operatorname{ord}_p(k)} \equiv 1 \pmod{p}$ and $\ldots k^{i\operatorname{ord}_p(k)} \equiv 1 \pmod{p}$ and no intermediate powers are congruent to 1 modulo p. Since $k^{p-1} \equiv 1 \pmod{p}$, then some multiple of $\operatorname{ord}_p(k)$ must equal $p-1$.

If you get in the habit of remembering sketches of proofs like the above every time you recall a theorem, then soon the proofs and the theorems will become much more real and immediate to you.

Returning now to the orders of elements modulo a prime p, we know that the order of every integer divides $p-1$. An integer whose order is as large as possible, namely $p-1$, has special significance, because, as you will soon prove, its powers give every non-zero member of a complete residue system modulo p. We first give such numbers a name and then prove that theorem.

Definition. Let p be a prime. An integer g such that $\operatorname{ord}_p(g) = p-1$ is called a *primitive root modulo p*.

6.6 Theorem. *Let p be a prime and suppose g is a primitive root modulo p. Then the set $\{0, g, g^2, g^3, \ldots, g^{p-1}\}$ forms a complete residue system modulo p.*

As usual, ideas become more meaningful if you look at actual numerical examples.

6.7 Exercise. *For each of the primes p less than 20 find a primitive root and make a chart showing what powers of the primitive root give each of the natural numbers less than p.*

Your exploration of the first few primes might suggest to you that every prime has at least one primitive root. In fact, that is true. We state that theorem here, and you may be able to think of a proof of it now; however, there are some preliminary theorems about the Euler ϕ-function that will help us to prove the existence of primitive roots. We will investigate those theorems in the next section and then return to this theorem about primitive roots.

6.8 Theorem. *Every prime p has a primitive root.*

One approach to proving the existence of primitive roots for a prime p is to put together a few of the ideas we already know. You proved that for any divisor d of $p - 1$, at most $\phi(d)$ incongruent numbers have order d modulo p. We know that every natural number k less than p has an order d that divides $p - 1$. So we could list the divisors d of $p - 1$ and for each such d we notice that at most $\phi(d)$ of the numbers $1, 2, 3, \ldots, p - 1$ have order d and systematically cross the order d numbers off the list. Let's try this strategy with the prime $p = 13$.

6.9 Exercise. *Consider the prime $p = 13$. For each divisor $d = 1, 2, 3,$ $4, 6, 12$ of $12 = p - 1$, mark which of the natural numbers in the set $\{1, 2, 3, 4, 5, 6, 7, 8, 9, 10, 11, 12\}$ have order d.*

Notice in the above exercise that there are $\phi(d)$ numbers of order d for each d. Of course, each number from 1 to 12 has some order. So in the case of 12,

$$\phi(1) + \phi(2) + \phi(3) + \phi(4) + \phi(6) + \phi(12) = 12.$$

A more compact way of writing the above sum is to use summation notation. We will write

$$\sum_{d \mid n} \phi(d)$$

for the sum of the Euler ϕ-function of the natural number divisors of the natural number n. So, for example, the previous observation can be written

$$\sum_{d \mid 12} \phi(d) = 12.$$

This example is suggestive of a more general relationship between the Euler ϕ-function and the divisors of a natural number, which we will explore in the next section.

Euler's ϕ-function and sums of divisors

For the moment, let's not think about primes and primitive roots and instead just look at any natural numbers. The first exercise below asks you to look at all the natural number divisors of a natural number, take the Euler ϕ-function of each divisor, add up those values and look for a pattern.

6.10 Exercise. *Compute each of the following sums.*

1. $\displaystyle\sum_{d|6} \phi(d)$

2. $\displaystyle\sum_{d|10} \phi(d)$

3. $\displaystyle\sum_{d|24} \phi(d)$

4. $\displaystyle\sum_{d|36} \phi(d)$

5. $\displaystyle\sum_{d|27} \phi(d)$

Make a sweeping conjecture about the sum of $\phi(d)$ taken over all the natural number divisors of any natural number n.

Your sweeping conjecture is probably true. To be sure, check Theorem 6.15 below. Since every natural number larger than 1 is the product of primes, we adopt the strategy of seeing how to prove the conjecture for primes and then seeing how to compute it for products of primes. In the case of primes, there are not many divisors to consider, so that simplifies the situation.

6.11 Lemma. *If p is a prime, then*

$$\sum_{d|p} \phi(d) = p.$$

You can list all the divisors of powers of primes very specifically. So that is the next case to tackle.

6.12 Lemma. *If p is a prime, then*

$$\sum_{d|p^k} \phi(d) = p^k.$$

To build up our understanding, the easiest case that involves more than one prime would be a natural number that is the product of exactly two primes. So that is the next case that we ask you to prove.

6.13 Lemma. *If p and q are two different primes, then*

$$\sum_{d|pq} \phi(d) = pq.$$

The proof of the preceding lemma has allowed you to develop the insights that enable you to deal with the product of any two relatively prime natural numbers, which is what you will do next.

6.14 Lemma. *If n and m are relatively prime natural numbers, then*

$$\left(\sum_{d|m} \phi(d)\right) \cdot \left(\sum_{d|n} \phi(d)\right) = \sum_{d|mn} \phi(d).$$

All the preceding lemmas allow you to finally prove your conjecture that the sum

$$\sum_{d|n} \phi(d)$$

will just equal the natural number that you started with.

6.15 Theorem. *If n is a natural number, then*

$$\sum_{d|n} \phi(d) = n.$$

After thinking about an idea for a few hundred years, it is sometimes possible to see the same result from a different point of view. The approach above is a clear strategy of doing simpler cases first and putting them together to get the result. But in this case, there is a slick alternative proof to the above theorem, which we thought you might enjoy. So please verify the steps of the following different approach to the same theorem.

6.16 Exercise. *For a natural number n consider the fractions*

$$\frac{1}{n}, \frac{2}{n}, \frac{3}{n}, \dots, \frac{n}{n},$$

all written in reduced form. For example, with n = 10 we would have

$$\frac{1}{10}, \frac{1}{5}, \frac{3}{10}, \frac{2}{5}, \frac{1}{2}, \frac{3}{5}, \frac{7}{10}, \frac{4}{5}, \frac{9}{10}, \frac{1}{1}.$$

Try to find a natural one-to-one correspondence between the reduced fractions and the numbers $\phi(d)$ for $d \mid n$. Show how that observation provides a very clever proof to the preceding theorem.

Having established the theorem that

$$\sum_{d \mid n} \phi(d) = n,$$

we can now prove that every prime p has a primitive root. In fact, we can prove that it has $\phi(p-1)$ primitive roots.

6.17 Theorem. *Every prime p has $\phi(p-1)$ primitive roots.*

Euler's ϕ-function is multiplicative

Although we defined the Euler ϕ-function, saw how to use it to prove a generalization of Fermat's Little Theorem, and saw how it was used in the discussion of primitive roots, we do not yet know how to compute the value of the Euler ϕ-function for an arbitrary natural number n. Since every natural number larger than 1 is the product of primes, we adopt the strategy of seeing how to compute the Euler ϕ-function for primes and then we see how to compute it for products of primes. We'll first ask you to make and prove a conjecture about the value of the Euler ϕ-function of a prime.

6.18 Exercise. *Make a conjecture about the value $\phi(p)$ for a prime p. Prove your conjecture.*

The next simpler kind of natural number is a product of primes where just one prime is involved, in other words, a power of a prime. Once again, we ask you to make a conjecture and prove it about the value of the Euler ϕ-function for powers of primes. If you get stuck, try just writing out the natural numbers $1, 2, 3, 4, ..., p^k$ for some primes p and small powers k and just circle those numbers on the list that are relatively prime to p^k. By looking at examples and looking for patterns, you can make and prove your conjecture for a formula that tells us $\phi(p^k)$.

6.19 Exercise. *Make a conjecture about the value $\phi(p^k)$ for a prime p and natural numbers k. Prove your conjecture.*

Our goal is to be able to compute the Euler ϕ-function for any natural number n. To do so, we first observe that the Euler ϕ-function counts relatively prime members of any complete residue system. That is, the Euler ϕ-function $\phi(n)$ counts the number of numbers in the set $\{1, 2, 3, \ldots, n\}$ that are relatively prime to n, but it also counts the number of numbers in any complete residue system modulo n that are relatively prime to n.

6.20 Theorem. *If n is a natural number and A is a complete residue system modulo n, then the number of numbers in A that are relatively prime to n is equal to $\phi(n)$.*

We can construct a complete residue system for a natural number n by taking an arithmetic progression of numbers where the steps are relatively prime to n.

6.21 Theorem. *If n is a natural number, k is an integer, and m is an integer relatively prime to n, then the set of n integers*

$$\{k, k + m, k + 2m, k + 3m, \ldots, k + (n - 1)m\}$$

is a complete residue system modulo n.

The previous two theorems can be used to prove the next theorem which states that the Euler ϕ-function of a product of relatively prime numbers is equal to the product of the Euler ϕ-functions of each. You might gain some insight by taking a few examples of relatively prime natural numbers m and n.

6.22 Exercise. *Consider the relatively prime natural numbers 9 and 4. Write down all the natural numbers less than or equal to $36 = 9 \cdot 4$ in a rectangular array that is 9 wide and 4 high. Then circle those numbers in that array that are relatively prime to 36. Try some other examples using relatively prime natural numbers.*

Now, using the insights you have gained from the examples, prove the following theorem.

6.23 Theorem. *If n and m are relatively prime natural numbers, then*

$$\phi(mn) = \phi(m)\phi(n).$$

Definition. A function f of natural numbers is *multiplicative* if and only if for any pair of relatively prime natural numbers m and n, $f(mn) = f(m)f(n)$.

The previous theorem could be restated by saying that the Euler ϕ-function is multiplicative. There are many other useful and interesting multiplicative functions in number theory, none of which will appear in this book.

We can now compute the Euler ϕ-function of any natural number by taking its unique prime factorization.

6.24 Exercise. *Compute each of the following.*

1. $\phi(3)$

2. $\phi(5)$

3. $\phi(15)$

4. $\phi(45)$

5. $\phi(98)$

6. $\phi(5^6 11^4 17^{10})$

We can now be more specific about what powers of numbers will be congruent to 1 modulo n.

6.25 Question. *To what power would you raise* 15 *to be certain that you would get an answer that is congruent to* 1 *modulo* 98? *Why?*

We can now compute the number of primitive roots of a prime.

6.26 Question. *How many primitive roots does the prime* 251 *have?*

Roots modulo a number

In Chapter 4 we investigated the process of repeated multiplication of numbers modulo another number, that is, taking powers of numbers and reducing those powers modulo n. Finding a number that when multiplied by itself k times results in the number b modulo n translates into solving the congruence

$$x^k \equiv b \pmod{n}.$$

A solution could be called a kth root of b modulo n. Our work on orders of elements and primitive roots sheds some light on the nature of the set of solutions when n is a prime and $b = 1$. Finding general solutions to congruences of this form is a difficult task to accomplish, but for certain choices of k, b, and n success is within our grasp.

Our goal is to develop a technique using Euler's Theorem for finding solutions to congruences of the form $x^k \equiv b$ (mod n), that is, finding kth roots of b modulo a number n. You have already seen instances of this technique in Chapter 5. Let's begin by experimenting with actual numbers.

6.27 Exercise. *Try, using paper and pencil, to solve several congruences of the form $x^k \equiv b$ (mod 5) and $x^k \equiv b$ (mod 6).*

We hope you observed that depending on the choice of k, b, and n in the previous exercise the congruence may have no solutions, one solution, or more than one solution. (If you did not observe this go try more examples!) In the next exercise you are asked to make an observation (one that you may very well have made already) that will get us on track for developing a more systematic strategy for finding kth roots modulo n.

6.28 Exercise. *Compute a^9 (mod 5) for several choices of a. Can you explain what happens? Now compute a^{17} (mod 15) for several choices of a. Does your previous explanation apply here too?*

The following theorem should capture your explanations from the last exercise. It is a straightforward and hopefully enlightening consequence of Euler's Theorem.

6.29 Theorem. *If a is an integer and v and n are natural numbers such that $(a, n) = 1$, then $a^{v\phi(n)+1} \equiv a$ (mod n).*

Now let's apply these observations to solve actual congruences.

6.30 Question. *Consider the congruence $x^5 \equiv 2$ (mod 7). Can you think of an appropriate operation we can apply to both sides of the congruence that would allow us to "solve" for x? If so, is the value obtained for x a solution to the original congruence?*

6.31 Question. *Consider the congruence $x^3 \equiv 7$ (mod 10). Can you think of an appropriate operation we can apply to both sides of the congruence that would allow us to "solve" for x? If so, is the value obtained for x a solution to the original congruence?*

We hope you discovered that raising both sides of our congruence to an appropriately chosen exponent seems to always yield a solution. The following theorem, which generalizes Theorem 5.3, asserts that such an exponent is always available.

6.32 Theorem. *If k and n are natural numbers with $(k, \phi(n)) = 1$, then there exist positive integers u and v satisfying $ku = \phi(n)v + 1$.*

The previous theorem not only asserts that an appropriate exponent is always available, but it also tells us how to find it. The numbers u and v are solutions to a linear Diophantine equation just like those we studied in Chapter 1.

6.33 Exercise. *Use your observations so far to find solutions to the following congruences. Be sure to check that your answers are indeed solutions.*

1. $x^7 \equiv 4 \pmod{11}$

2. $x^5 \equiv 11 \pmod{18}$

3. $x^7 \equiv 2 \pmod{8}$

You have probably devised a method for finding a solution to a congruence of the form $x^k \equiv b \pmod{n}$, but the third congruence in the above exercise shows that this method does not always work.

6.34 Question. *What hypotheses on k, b, and n do you think are necessary for your method to produce a solution to the congruence $x^k \equiv b \pmod{n}$? Make a conjecture and prove it.*

6.35 Theorem. *If b is an integer and k and n are natural numbers such that $(k, \phi(n)) = 1$ and $(b, n) = 1$, then $x^k \equiv b \pmod{n}$ has a unique solution modulo n. Moreover, that solution is given by*

$$x \equiv b^u \pmod{n},$$

where u and v are positive integers such that $ku = \phi(n)v + 1$.

Our experiments at the beginning of the section showed that a number can have multiple roots modulo another number. But the previous theorem asserts that under the given hypotheses, our method not only finds a kth root modulo n, but in fact finds the *only* kth root.

6.36 Exercise. *Find the 49th root of 100 modulo 151.*

The following two theorems assert that for square-free numbers n, that is, numbers that are products of distinct primes, the hypothesis $(b, n) = 1$ from Theorem 6.35 can be dropped. The first theorem is a generalization of Theorem 5.2.

6.37 Theorem. *If a is an integer, v is a natural number, and n is a product of distinct primes, then* $a^{v\phi(n)+1} \equiv a$ (mod *n*).

6.38 Theorem. *If n is a natural number that is a product of distinct primes, and k is a natural number such that* $(k, \phi(n)) = 1$, *then* $x^k \equiv b$ (mod *n*) *has a unique solution modulo n for any integer b. Moreover, that solution is given by*

$$x \equiv b^u \pmod{n},$$

where u and v are positive integers such that $ku - \phi(n)v = 1$.

6.39 Exercise. *Find the 37th root of 100 modulo 210.*

General solutions to the congruence $x^k \equiv b$ (mod *n*) when $(k, \phi(n)) > 1$ are much harder to come by. In Chapter 7 we will consider in depth the special case of $k = 2$ and *n* a prime. Using our work on primitive roots modulo a prime we can prove the following final result which tells us something about the number of roots a number can have modulo a prime.

6.40 Theorem. *Let p be a prime, b an integer, and k a natural number. Then the number of kth roots of b modulo p is either 0 or* $(k, p-1)$.

6.41 Blank Paper Exercise. *After not looking at the material in this chapter for a day or two, take a blank piece of paper and outline the development of that material in as much detail as you can without referring to the text or to notes. Places where you get stuck or can't remember highlight areas that may call for further study.*

Sophie Germain is Germane, Part I

We hope your work so far has convinced you of the usefulness of primitive roots modulo a prime *p*. The powers of a primitive root produce a complete residue system that is often as useful as the canonical system. From a practical point of view, finding a primitive root is a necessary ingredient in the Diffie-Hellman public key exchange described in the last chapter. But although their existence is guaranteed, finding a primitive root modulo *p* is not completely straightforward.

We know that a prime *p* has $\phi(p-1)$ primitive roots, which can be a large proportion of the numbers modulo *p*. For example, the prime 65,537 has 37,768 primitive roots (although the preceeding prime 65,521 has only 13,824 primitive roots). So trial and error is likely to produce a primitive root without much trouble. But trial and error is an irksome procedure to many mathematicians. For them we offer the following theorem.

Theorem (A Primitive Root Test). *Let p be a prime. Then a is a primitive root modulo p if and only if for all factors f of $p-1$,*

$$a^{\frac{p-1}{f}} \not\equiv 1 \pmod{p}.$$

This test just asserts that if $\operatorname{ord}_p(a)$ is not a proper divisor of $p-1$, then a is a primitive root. But this is hardly a new insight. In addition, performing this test requires factoring $p-1$, which is one of our "hard problems."

Unfortunately we do not have a recipe for conjuring up a primitive root for an arbitrary prime. The mathematician Emil Artin (1898–1962) made a conjecture regarding primitive roots that would imply the following.

Conjecture (Artin's Conjecture). *Every integer which is neither -1 nor a perfect square is a primitive root for infinitely many primes.*

The conjecture is still unproven. In fact, there is not a single integer satisfying the hypotheses of Artin's Conjecture for which we know the conjecture to be true, although such a statement is not meant to imply that no progress has been made. For example, we know that it suffices to show that the conjecture is true for just the primes; that is, it suffices to show that every prime is a primitive root for infinitely many other primes.

Strangely, although we cannot cite a single example for which Artin's Conjecture is true, we know that there are no more than two exceptions. But we have no idea what those exceptions might be. So for example, it is known that at least one of the primes 3, 5, or 7 is a primitive root for infinitely many primes, but we can't say for sure that 3 is or that 5 is or that 7 is! It's also known that at least one of the primes 67867979, 256203221, or 2899999517 is a primitive root for infinitely many primes. If you are a betting person, we suggest you bet a dollar that 2899999517 is a primitive root for infinitely many primes. If you are ever proved wrong, we'll buy you a fancy dinner at the restaurant of your choice and a car.

Sometimes, focusing on primes of a special form can lead to interesting progress. Sophie Germain (1776–1831) was a French mathematician who made wonderful contributions to number theory. For cultural reasons of the period, she communicated much of her early work under the male pseudonym "Monsieur Le Blanc." Under this pseudonym, she submitted one of her early manuscripts to Lagrange. Aware of the mathematical talent required to produce such work, Lagrange discovered her true identity and became a mathematical mentor to Germain.

Sophie Germain is credited with making one of history's great advances towards a proof of Fermat's Last Theorem. Fermat's Last Theorem is the statement that there are no natural number solutions to the Diophantine equation

$$x^q + y^q = z^q$$

when q is a natural number greater than 2. Sophie Germain studied the famous Fermat equation $x^q + y^q = z^q$ for primes q with the property that $p = 2q + 1$ is also prime. Such primes are now known as *Sophie Germain primes*.

The orders of elements modulo a prime $p = 2q + 1$, where q is also prime, are very restricted. In fact, since the order of any element must divide $p - 1 = 2q$, we see that the only possible orders are 1, 2, q, and $2q$. There is only one element of order 1 (namely 1 itself), and only one element of order 2 (namely $p - 1$). And so the remaining elements split into those of order q and those of order $2q$, the latter being our primitive roots. In a 1909 paper titled *Methods to Determine the Primitive Root of a Number*, G. A. Miller showed there is at least one element we can always count on to be in this latter group.

Theorem (Miller's Theorem). *Let p be an odd prime of the form $p = 2q + 1$ where q is an odd prime. Then the complete set of primitive roots modulo p are $-(2)^2, -(3)^2, \ldots, -(q)^2$. In particular, -4 is a primitive root of every prime of this form.*

So why didn't Miller find the first example of an integer for which Artin's Conjecture holds? Alas, unfortunately, it is still unknown whether or not there are infinitely many Sophie Germain primes.

In the next chapter we introduce the Law of Quadratic Reciprocity, which will then allow you to prove Miller's theorem above and describe a satisfying symmetry among primitive roots and perfect squares modulo p in the world of Sophie Germain primes p.

7

The Golden Rule:
Quadratic Reciprocity

Quadratic Congruences

We previously analyzed the solutions to all linear Diophantine equations modulo a number n, that is, we investigated congruences $ax \equiv b \pmod{n}$. We proved that we can find at least one number x that satisfies that congruence if and only if $(a, n)|b$. Now we investigate quadratics modulo n, that is, congruences that involve an unknown that is squared. As always, our exploration of this question begins with the easiest case we can think of, namely where the modulus is a prime and the quadratic expression is just to square x. In other words, we want to understand the congruence

$$x^2 \equiv a \pmod{p},$$

where a is an integer and p is a prime. We seek to answer the question, "Which numbers are perfect squares modulo p and which are not?"

This exploration of perfect squares modulo a prime p has fascinating insights that attracted the attention of some of the greatest mathematicians of all time.

Quadratic residues

Our first two theorems assert that our simplest quadratic congruences actually encompass all cases. That is, any quadratic congruence modulo a prime can be replaced with a much simpler congruence.

7.1 Theorem. *Let p be a prime and let a, b, and c be integers with a not divisible by p. Then there are integers b' and c' such that the set of* *solutions to the congruence* $ax^2 + bx + c \equiv 0$ *(mod p) is equal to the set of solutions to a congruence of the form* $x^2 + b'x + c' \equiv 0$ *(mod p).*

7.2 Theorem. *Let p be a prime, and let b and c be integers. Then there exists a linear change of variable,* $y = x + \alpha$ *with α an integer, transforming the congruence* $x^2 + bx + c \equiv 0$ *(mod p) into a congruence of the form* $y^2 \equiv \beta$ *(mod p) for some integer β.*

Our goal is to understand which integers are perfect squares of other integers modulo a prime p. The first theorem below tells us that half the natural numbers less than an odd prime p are perfect squares and half are not. To prove that theorem and some of the others in the chapter, keep the idea of a primitive root in mind. Remember that every prime p has a primitive root g and the set $\{0, g^1, g^2, g^3, \ldots, g^{(p-1)}\}$ forms a complete residue system modulo p. This picture of the numbers modulo p is frequently valuable.

7.3 Theorem. *Let p be an odd prime. Then half the numbers not congruent to 0 in any complete residue system modulo p are perfect squares modulo p and half are not.*

As usual, it is a good idea to look at a specific example. You may want to do the following exercise with several primes.

7.4 Exercise. *Determine which of the numbers 1, 2, 3,..., 12 are perfect squares modulo 13. For each such perfect square, list the number or numbers in the set whose square is that number.*

The following question asks you to rephrase your insight about perfect squares modulo a prime p in terms of their representation as the power of a primitive root.

7.5 Question. *Can you characterize perfect squares modulo a prime p in terms of their representation as a power of a primitive root?*

Perfect squares modulo a prime p attracted so much interest from number theorists that such squares are given their own alternative name, quadratic residue. Here is the definition.

Definition. If a is an integer and p is a prime and $a \equiv b^2$ (mod p) for some integer b, then a is called a *quadratic residue modulo p*. If a is not congruent to any square modulo p, then a is a *quadratic non-residue modulo p*.

We can rephrase our previous theorem in terms of quadratic residues.

7.6 Theorem. *Let p be a prime. Then half the numbers not congruent to 0 modulo p in any complete residue system modulo p are quadratic residues modulo p and half are quadratic non-residues modulo p.*

From elementary school days, we have known that the product of a positive number and a positive number is positive, a positive times a negative is negative, and the product of two negative numbers is positive. Quadratic residues and non-residues are related similarly.

7.7 Theorem. *Suppose p is an odd prime and p does not divide either of the two integers a or b. Then*

1. *If a and b are both quadratic residues modulo p, then ab is a quadratic residue modulo p;*

2. *If a is a quadratic residue modulo p and b is a quadratic non-residue modulo p, then ab is a quadratic non-residue modulo p;*

3. *If a and b are both quadratic non-residues modulo p, then ab is a quadratic residue modulo p.*

One of the mathematicians who studied quadratic residues modulo p was the French mathematician Legendre. He invented a symbol called the Legendre symbol that gives a value of 1 to quadratic residues and -1 to quadratic non-residues. The symbol is convenient because it lets us express theorems like the previous one in a compact way. Here is the definition.

Definition. For an odd prime p and a natural number a with p not dividing a, the *Legendre symbol* $\left(\frac{a}{p}\right)$ is defined by

$$\left(\frac{a}{p}\right) = \begin{cases} 1 & \text{if } a \text{ is a quadratic residue modulo } p, \\ -1 & \text{if } a \text{ is a quadratic non-residue modulo } p. \end{cases}$$

Now we can express the preceding theorem using the Legendre symbol.

7.8 Theorem. *Suppose p is an odd prime and p does not divide either a or b. Then*

$$\left(\frac{ab}{p}\right) = \left(\frac{a}{p}\right)\left(\frac{b}{p}\right).$$

Our goal is to be able to take an integer a and determine whether it is a quadratic residue modulo a prime p or a quadratic non-residue. Euler gave one method for determining whether a number is a quadratic residue. The method depends on raising the number to the $(p-1)/2$ power.

7.9 Theorem (Euler's Criterion). *Suppose p is an odd prime and p does not divide the natural number a. Then a is a quadratic residue modulo p if and only if $a^{(p-1)/2} \equiv 1 \pmod{p}$; and a is a quadratic non-residue modulo p if and only if $a^{(p-1)/2} \equiv -1 \pmod{p}$. This criterion can be abbreviated using the Legendre symbol:*

$$a^{(p-1)/2} \equiv \left(\frac{a}{p}\right) \pmod{p}.$$

The number 1 is always a quadratic residue. Other numbers modulo p sometimes are and sometimes are not quadratic residues, depending on p, but we can give a good description for when a number congruent to -1 modulo a prime p is a quadratic residue.

7.10 Theorem. *Let p be an odd prime. Then -1 is a quadratic residue modulo p if and only if p is of the form $4k+1$ for some integer k. Or, equivalently,*

$$\left(\frac{-1}{p}\right) = \begin{cases} 1 & \text{if } p \equiv 1 \pmod{4}, \\ -1 & \text{if } p \equiv 3 \pmod{4}. \end{cases}$$

The following theorem identifies the square roots of -1 modulo p when p is congruent to 1 modulo 4.

7.11 Theorem. *Let k be a natural number and $p = 4k+1$ be a prime congruent to 1 modulo 4. Then*

$$(\pm(2k)!)^2 \equiv -1 \pmod{p}.$$

We end this section with one final application of Theorem 7.10. In Chapter 2 you proved there are infinitely many primes. Except for the prime 2, all primes are congruent to either 1 or 3 modulo 4. You proved that infinitely many primes are congruent to 3 modulo 4, but probably did not show that infinitely many primes are congruent to 1 modulo 4.

7.12 Theorem (Infinitude of $4k+1$ Primes Theorem). *There are infinitely many primes congruent to 1 modulo 4.*
    (Hint: If $p_1, p_2, \ldots, p_r$ are primes each congruent to 1 modulo 4, what can you say about each prime factor of the number $N = (2p_1 p_2 \cdots p_r)^2 + 1$?)

Gauss' Lemma and quadratic reciprocity

Euler's criterion worked well for analyzing whether or not -1 is a quadratic residue or quadratic non-residue. But the computation of $a^{(p-1)/2}$ modulo p for a general value of a is a non-trivial task. Gauss gave us a useful lemma which will allow us to proceed a little further with our strategy of analyzing particular numbers.

It will be useful to have in mind a proof strategy that we found useful for proving Fermat's Little Theorem and Euler's Theorem. One proof of Fermat's Little Theorem involved multiplying $1a \cdot 2a \cdot 3a \cdots (p-1)a$ and gathering the a's to get the factor $a^{(p-1)}$. Euler's Criterion tells us that an integer a relatively prime to p is a quadratic residue modulo p if and only if $a^{(p-1)/2}$ is congruent to 1 modulo p. So let's think about producing a product that will have $a^{(p-1)/2}$ in it.

Consider the numbers a, $2a$, $3a$, ..., $\frac{p-1}{2}a$ modulo p. These numbers are distinct modulo p (do you recall why?) and each is congruent to a member of the complete residue system

$$\left\{ -\frac{p-1}{2}, \ldots, -1, 0, 1, \ldots, \frac{p-1}{2} \right\}.$$

The product of these numbers, regardless of which complete residue system they come from, are congruent modulo p.

For example, consider the case of $a = 3$ and $p = 11$, so $\frac{(p-1)}{2} = 5$. We obtain the congruence

$$3 \cdot 2(3) \cdot 3(3) \cdot 4(3) \cdot 5(3) \equiv 3 \cdot -5 \cdot -2 \cdot 1 \cdot 4 \quad (\text{mod } 11),$$

or

$$3^5 \cdot 5! \equiv 5! \quad (\text{mod } 11).$$

Since $5!$ is not divisible by 11 we may cancel it from both sides to obtain $3^5 \equiv 1 \pmod{11}$ which, by Euler's Criterion, tells us that 3 is a quadratic residue modulo 11. The following lemma asserts that it was not just a coincidence that we obtained $5!$ on both sides of the congruence.

7.13 Lemma. *Let p be a prime, a an integer not divisible by p, and r_1, r_2, ..., $r_{\frac{(p-1)}{2}}$ the representatives of a, $2a$, ..., $\frac{p-1}{2}a$ in the complete residue system*

$$\left\{ -\frac{p-1}{2}, \ldots, -1, 0, 1, \ldots, \frac{p-1}{2} \right\}.$$

Then

$$r_1 \cdot r_2 \cdots \cdots r_{\frac{(p-1)}{2}} = (-1)^g \left(\frac{p-1}{2} \right)!,$$

where g is the number of r_i's which are negative.

(Hint: It suffices to show that we never have $r_i \equiv -r_j \pmod{p}$ for some i and j.)

7.14 Theorem (Gauss' Lemma). *Let p be a prime and a an integer not divisible by p. Let g be the number of negative representatives of a, $2a$, ..., $\frac{p-1}{2}a$ in the complete residue system $\left\{ -\frac{p-1}{2}, \ldots, -1, 0, 1, \ldots, \frac{p-1}{2} \right\}$. Then*

$$\left(\frac{a}{p} \right) = (-1)^g.$$

We now apply Gauss' Lemma to characterize those primes p for which 2 is a quadratic residue. Consider the following data. Notice that for some primes p, 2 is a quadratic residue modulo p, and for others it is not. Make a conjecture that characterizes the primes in each category. The question after the data gives you a hint, so you might enjoy trying to devise your characterization before looking at the next question.

Here are the first primes for which 2 is a quadratic residue:

$$7, 17, 23, 31, 41, 47, 71, 73, 79, 89, 97, 103, 113, 127.$$

Here are the first primes for which 2 is a quadratic non-residue:

$$3, 5, 11, 13, 19, 29, 37, 43, 53, 59, 61, 67, 83, 101, 107, 109.$$

7.15 Question. *Does the prime's residue class modulo 4 determine whether or not 2 is a quadratic residue? Consider the primes' residue class modulo 8 and see whether the residue class seems to correlate with whether or not 2 is a quadratic residue. Make a conjecture.*

7.16 Theorem. *Let p be an odd prime, then*

$$\left(\frac{2}{p} \right) = \begin{cases} 1 & \text{if } p \equiv 1 \text{ or } 7 \pmod{8}, \\ -1 & \text{if } p \equiv 3 \text{ or } 5 \pmod{8}. \end{cases}$$

You might fear that we will proceed to analyze $\left(\frac{3}{p} \right)$, then $\left(\frac{4}{p} \right)$, then $\left(\frac{5}{p} \right)$, and so on for ever; however, fortunately, there is a shortcut. The shortcut occurs by making an observation about pairs of primes. If you have

7. The Golden Rule: Quadratic Reciprocity

two odd primes p and q, then whether p is a quadratic residue modulo q and whether q is a quadratic residue modulo p are related. As we have seen in the cases of -1 and 2, questions of being a perfect square modulo p are related to what p is modulo 4 or 8, so it is natural to consider the residues of p and q modulo 4 while investigating the relationship between $\left(\frac{p}{q}\right)$ and $\left(\frac{q}{p}\right)$.

7.17 Exercise. *Table 1 shows* $\left(\frac{p}{q}\right)$ *for the first several odd primes. For example, the table indicates that* $\left(\frac{7}{3}\right) = 1$, *but that* $\left(\frac{3}{7}\right) = -1$. *Make another table that shows when* $\left(\frac{p}{q}\right) = \left(\frac{q}{p}\right)$ *and when* $\left(\frac{p}{q}\right) \neq \left(\frac{q}{p}\right)$.

	3	5	7	11	13	17	19	23	29	31	37	41	43	47
3		-1	1	-1	1	-1	1	-1	-1	1	1	-1	1	-1
5	-1		-1	1	-1	-1	1	-1	1	1	-1	1	-1	-1
7	-1	-1		1	-1	-1	-1	1	1	-1	1	-1	1	-1
11	1	1	-1		-1	-1	-1	1	-1	1	1	-1	-1	1
13	1	-1	-1	-1		1	-1	1	1	-1	-1	-1	1	-1
17	-1	-1	-1	-1	1		1	-1	-1	-1	-1	-1	1	1
19	-1	1	1	1	-1	1		1	-1	-1	-1	-1	1	1
23	1	-1	-1	-1	1	-1	-1		1	1	-1	1	-1	1
29	-1	1	1	-1	1	-1	-1	1		-1	-1	-1	-1	-1
31	-1	1	1	-1	-1	-1	1	-1	-1		-1	1	-1	1
37	1	-1	1	1	-1	-1	-1	-1	-1	-1		1	-1	1
41	1	1	-1	-1	-1	-1	-1	1	-1	1	1		1	-1
43	-1	-1	-1	1	1	1	-1	1	-1	1	-1	1		1
47	1	-1	1	-1	-1	1	-1	-1	-1	-1	1	-1	-1	

Table 1. Values of $\left(\frac{p}{q}\right)$ for p across the top and q down the side.

7.18 Exercise. *Make a conjecture about the relationship between* $\left(\frac{p}{q}\right)$ *and* $\left(\frac{q}{p}\right)$ *depending on* p *and* q.

Your conjecture is called "quadratic reciprocity."

7.19 Theorem (Quadratic Reciprocity Theorem—Reciprocity Part). *Let p and q be odd primes, then*

$$\left(\frac{p}{q}\right) = \begin{cases} \left(\frac{q}{p}\right) & \text{if } p \equiv 1 \pmod 4 \text{ or } q \equiv 1 \pmod 4, \\ -\left(\frac{q}{p}\right) & \text{if } p \equiv q \equiv 3 \pmod 4. \end{cases}$$

(Hint: Try to use the techniques used in the case of $\left(\frac{2}{p}\right)$.)

Putting together all our insights, we can write one theorem that we call the Law of Quadratic Reciprocity.

Theorem (Law of Quadratic Reciprocity). *Let p and q be odd primes, then*

1. $\left(\dfrac{-1}{p}\right) = \begin{cases} 1 & \text{if } p \equiv 1 \pmod 4, \\ -1 & \text{if } p \equiv 3 \pmod 4, \end{cases}$

2. $\left(\dfrac{2}{p}\right) = \begin{cases} 1 & \text{if } p \equiv 1 \pmod 8 \text{ or } p \equiv 7 \pmod 8, \\ -1 & \text{if } p \equiv 3 \pmod 8 \text{ or } p \equiv 5 \pmod 8, \end{cases}$

3. $\left(\dfrac{p}{q}\right) = \begin{cases} \left(\frac{q}{p}\right) & \text{if } p \equiv 1 \pmod 4 \text{ or } q \equiv 1 \pmod 4, \\ -\left(\frac{q}{p}\right) & \text{if } p \equiv q \equiv 3 \pmod 4. \end{cases}$

Recall that we proved that if p is an odd prime and p does not divide a or b, then $\left(\frac{ab}{p}\right) = \left(\frac{a}{p}\right)\left(\frac{b}{p}\right)$. That fact along with the Law of Quadratic Reciprocity lets us develop an effective technique for determining for any integer a whether or not it is a quadratic residue modulo the prime p.

7.20 Exercise (Computational Technique). *Given a prime p, show how you can determine whether a number a is a quadratic residue modulo p. Equivalently, show how to find $\left(\frac{a}{p}\right)$. To illustrate your method, compute $\left(\frac{1248}{93}\right)$ and some other examples.*

7.21 Exercise. *Find all the quadratic residues modulo 23.*

The Law of Quadratic Reciprocity allows us to determine whether or not an integer is a perfect square modulo a prime p; however, it does not help us to actually find the square roots. Sometimes we can obtain general expressions for certain square roots, as we did in Theorem 7.11. But there is no known algorithm for doing that in general.

Sophie Germain is germane, Part II

Recall from Chapter 6 that a *Sophie Germain prime* is a prime q for which $p = 2q + 1$ is also prime. For example, 23 is a Sophie Germain prime since $47 = 2 \cdot 23 + 1$ is also prime.

We know that for any prime p, the order of any integer a relatively prime to p must divide $p - 1$. If p is a prime, $p = 2q + 1$, and q is also prime, then $p - 1 = 2q$, so the elements modulo p have orders either 1, 2, q, or $2q$ (since these are all the possible divisors of $p - 1$). We know that 1 and $p - 1$ are, respectively, the only elements of order 1 and 2. So we conclude that every natural number a with $1 < a < p - 1$, must have order either q or $2q$, where those with order $2q$ are the primitive roots modulo p. Euler's Criterion can help us characterize the elements of order q.

7.22 Theorem. *Let p be a prime of the form $p = 2q + 1$ where q is a prime. Then every natural number a, $0 < a < p - 1$, is either a quadratic residue or a primitive root modulo p.*

Let's illustrate the above theorem by looking at the example furnished by the primes $q = 11$ and $p = 23$. According to the above theorem, each of the numbers 2, 3, ..., 21 is either a quadratic residue of order 11 ($= q$) modulo 23 or a primitive root modulo 23. In Exercise 7.21 you computed the quadratic residues modulo 23, yielding the numbers

$$2, 3, 4, 6, 8, 9, 12, 13, 16, 18 \quad (\text{mod } 23)$$

(the number 1 is a quadratic residue as well, but is not one of order q). It follows that the primitive roots modulo 23 must be given by

$$5, 7, 10, 11, 14, 15, 17, 19, 20, 21 \quad (\text{mod } 23).$$

And in fact, putting together the list of primitive roots (in bold) and the list of quadratic residues greater than 1 (underlined), we have

$$\underline{2}, \underline{3}, \underline{4}, \mathbf{5}, \underline{6}, \mathbf{7}, \underline{8}, \underline{9}, \mathbf{10}, \mathbf{11}, \underline{12}, \underline{13}, \mathbf{14}, \mathbf{15}, \underline{16}, \mathbf{17}, \underline{18}, \mathbf{19}, \mathbf{20}, \mathbf{21} \quad (\text{mod } 23),$$

which is a complete list of all numbers from 2 to 21 modulo 23.

A second and more subtle observation we might make about the above list of numbers modulo 23 has to do with symmetry. If you imagine a vertical line dividing the list between the numbers 11 and 12, a certain sort of mirror symmetry appears. In fact, it might be better described as "anti-symmetry", as the mirror image of a primitive root is a quadratic residue, and vice versa. This symmetry is a consequence of a more general property shared by primes arising from odd Sophie Germain primes.

7.23 Theorem. *Let p be a prime congruent to 3 modulo 4. Let a be a natural number with $1 < a < p - 1$. Then a is a quadratic residue modulo p if and only if $p - a$ is a quadratic non-residue modulo p.*

7.24 Theorem. *Let p be a prime of the form $p = 2q + 1$ where q is an odd prime. Then $p \equiv 3 \pmod 4$.*

The next theorem describes the symmetry between primitive roots and quadratic residues for primes arising from odd Sophie Germain primes.

7.25 Theorem. *Let p be a prime of the form $p = 2q + 1$ where q is an odd prime. Let a be a natural number, $1 < a < p - 1$. Then a is a quadratic residue if and only of $p - a$ is a primitive root modulo p.*

An attractive property of primes that arise from Sophie Germain primes is that they have primitive roots that we can actually compute. We saw the statement of this fact in Miller's Theorem in Chapter 6. Here we ask you to prove some theorems that will allow you to prove Miller's Theorem. We first note that perfect squares cannot be primitive roots modulo p for any prime p.

7.26 Theorem. *Let p be a prime and a be an integer. Then a^2 is not a primitive root modulo p.*

Next we see that natural numbers less than half a prime p cannot yield equivalent squares modulo p.

7.27 Theorem. *Let p be a prime and let i and j be natural numbers with $i \neq j$ satisfying $1 < i, j < \frac{p}{2}$. Then $i^2 \not\equiv j^2 \pmod p$.*

Now we start to deal with primes p that arise from Sophie Germain primes. Here we list all the integers modulo p that are not primitive roots modulo p.

7.28 Theorem. *Let p be a prime of the form $p = 2q + 1$ where q is an odd prime. Then the complete set of numbers that are not primitive roots modulo p are $1, -1, 2^2, 3^2, \ldots, q^2$.*

Now we can prove Miller's Theorem that characterizes the primitive roots of a prime that arises from a Sophie Germain prime.

7.29 Theorem. *Let p be a prime of the form $p = 2q + 1$ where q is an odd prime. Then the complete set of primitive roots modulo p are $-2^2, -3^2, \ldots, -q^2$.*

7.30 Exercise. *Verify that the primitive roots modulo 23 that we listed earlier in this section are in fact the same as those given by Miller's Theorem.*

7.31 Exercise. *List the primitive roots and quadratic residues modulo 47.*

We are able to analyze primes that arise from Sophie Germain primes successfully because we have such useful information about the prime factorization of $p - 1$. Of course, these special primes are rare. So many questions remain about how to find and describe primitive roots and perfect squares modulo more general primes.

7.32 Blank Paper Exercise. *After not looking at the material in this chapter for a day or two, take a blank piece of paper and outline the development of that material in as much detail as you can without referring to the text or to notes. Places where you get stuck or can't remember highlight areas that may call for further study.*

8

Pythagorean Triples, Sums of Squares, and Fermat's Last Theorem

Congruences to Equations

The Law of Quadratic Reciprocity gives us a neat view of which numbers are squares modulo a prime p. Information about squares modulo p can help us to understand actual numbers and equations in addition to modular numbers and congruences. In this chapter and the next we turn from quadratic congruences to quadratic (and higher order) Diophantine equations. We start with a quadratic equation we should all have some familiarity with from its connections to right triangles and the Pythagorean Theorem. Some of the questions will lead us to ask which numbers can be written as sums of squares, and the Law of Quadratic Reciprocity will help us find an answer. Finally, we turn to one of the most famous recent results of number theory, Fermat's Last Theorem.

Pythagorean triples

The Pythagorean Theorem asserts that the sum of the squares on the legs of a right triangle equals the square on the hypotenuse. Said another way, the lengths of the sides of a right triangle always provide a solution to the equation

$$x^2 + y^2 = z^2$$

by substituting the lengths of the legs for x and y and the length of the hypotenuse for z. In this section we consider the above quadratic as a Diophantine equation, that is, we consider only its integer solutions.

Definition. A triple of three positive integers (a, b, c) satisfying $a^2 + b^2 = c^2$ is called a *Pythagorean triple*.

Due to the close relationship with right triangles, the values a and b in a Pythagorean triple will sometimes be referred to as the *legs*, and the value c as the *hypotenuse*.

There are no Pythagorean triples in which both legs are odd.

8.1 Theorem. *If (a, b, c) is a Pythagorean triple, then at least one of a or b is even.*

The most famous Pythagorean triples are $(3, 4, 5)$ and $(5, 12, 13)$, but there are infinitely many. Let's begin by just finding a few.

8.2 Exercise. *Find at least seven different Pythagorean triples. Make a note of your methods.*

You may have discovered how to generate new Pythagorean triples from old ones through multiplication. Namely, if (a, b, c) is any Pythagorean triple and d is any natural number, then (da, db, dc) is also a Pythagorean triple. Pythagorean triples that are not simply multiples of smaller Pythagorean triples have a special designation.

Definition. A Pythagorean triple (a, b, c) is said to be *primitive* if a, b, and c have no common factor.

There are infinitely many primitive Pythagorean triples, so let's start by finding a few.

8.3 Exercise. *Find at least five primitive Pythagorean triples.*

We saw earlier that no Pythagorean triple has both legs odd, but for primitive Pythagorean triples, the legs cannot both be even either.

8.4 Theorem. *In any primitive Pythagorean triple, one leg is odd, one leg is even, and the hypotenuse is odd.*

It turns out that there is a method for generating infinitely many Pythagorean Triples in an easy way. It comes from looking at some simple algebra from high school. Remember that

$$(x + y)^2 = x^2 + 2xy + y^2$$

and

$$(x - y)^2 = x^2 - 2xy + y^2.$$

The difference between the two is $4xy$. So we have a relationship that looks almost like a Pythagorean triple, namely, one square $(x + y)^2$ equals another square $(x - y)^2$ plus something that we wish were a square, namely $4xy$. How could we ensure that $4xy$ is a square? Simple, just choose x and y to be squares. This kind of analysis leads to the following theorem.

8.5 Theorem. *Let s and t be any two different natural numbers with $s > t$. Then*

$$(2st, (s^2 - t^2), (s^2 + t^2))$$

is a Pythagorean triple.

The preceding theorem lets us easily generate infinitely many Pythagorean triples, but, in fact, every primitive Pythagorean triple can be generated by choosing appropriate natural numbers s and t and making the Pythagorean triple as described in the preceding theorem. As a hint to the proof, we make a little observation.

8.6 Lemma. *Let (a, b, c) be a primitive Pythagorean triple where a is the even number. Then $\frac{c+b}{2}$ and $\frac{c-b}{2}$ are perfect squares, say, s^2 and t^2, respectively; and s and t are relatively prime.*

So now we can completely characterize all primitive Pythagorean triples.

8.7 Theorem (Pythagorean Triple Theorem). *Let (a, b, c) be a triple of natural numbers with a even, b odd, and c odd. Then (a, b, c) is a primitive Pythagorean triple if and only if there exist relatively prime positive integers s and t, one even and one odd, such that $a = 2st$, $b = (s^2 - t^2)$, and $c = (s^2 + t^2)$.*

The formulas given in the Pythagorean Triple Theorem allow us to investigate the types of numbers that can occur in Pythagorean triples. Let's start our investigation by looking at examples.

8.8 Exercise. *Using the above formulas make a lengthy list of primitive Pythagorean triples.*

We'll begin by looking at the legs and then think about the hypotenuse later.

8.9 Exercise. *Make a conjecture that describes those natural numbers that can appear as legs in a primitive Pythagorean triple.*

You might have come up with the following theorem.

8.10 Theorem. *In every primitive Pythagorean triple, one leg is an odd integer greater than* 1 *and the other is a positive multiple of* 4.

This observation does not tell us which odd numbers are allowable or which multiples of 4 occur, but in fact every odd number and every multiple of 4 occurs as a leg in a Pythagorean triple.

8.11 Theorem. *Any odd number greater than* 1 *can occur as a leg in a primitive Pythagorean triple.*

8.12 Theorem. *Any positive multiple of* 4 *can occur as a leg in a primitive Pythagorean triple.*

To analyze what numbers can occur as the hypotenuse of a primitive Pythagorean triple is a bit trickier. It amounts to investigating the general problem of representing numbers as sums of two squares.

Sums of squares

The question we seek to answer is, for which numbers n does the Diophantine equation

$$x^2 + y^2 = n$$

have a solution? As usual we will first investigate the case of primes.

8.13 Question. *Make a list of the first fifteen primes and write each as the sum of as few squares of natural numbers as possible. Which ones can be written as the sum of two squares? Make a conjecture about which primes can be written as the sum of two squares of natural numbers.*

Your conjecture likely singles out those primes that are congruent to 1 modulo 4.

Theorem. *Let p be a prime. Then p can be written as the sum of two squares of natural numbers if and only if $p = 2$ or $p \equiv 1$ (mod 4).*

There are really two theorems here and we will state them separately below. The first is a much simpler theorem to prove than the second.

8.14 Theorem. *Let p be a prime such that $p = a^2 + b^2$ for some natural numbers a and b. Then either $p = 2$ or $p \equiv 1$ (mod 4).*

The fact that every prime congruent to 1 modulo 4 is expressible as the sum of two squares is more challenging to prove. As you work to prove this result in the next few theorems it is worthwhile to recall another theorem you recently proved about primes that are congruent to 1 modulo 4. For primes congruent to 1 modulo 4, -1 is a quadratic residue; that is, for any prime p that is congruent to 1 modulo 4, there is some natural number a such that a^2 is congruent to -1 modulo p. To prove the second theorem, try applying the following lemma to a square root of -1 modulo p.

8.15 Lemma. *Let p be a prime and let a be a natural number not divisible by p. Then there exist integers x and y such that $ax \equiv y \pmod{p}$ with $0 < |x|, |y| < \sqrt{p}$.*

8.16 Theorem. *Let p be a prime such that $p \equiv 1 \pmod 4$. Then p is equal to the sum of two squares of natural numbers.*
    (Hint: Try applying the previous lemma to a square root of -1 modulo p.)

Knowing which primes can be written as the sum of two squares is a great start, but that does not yet answer the question as to which numbers can occur as the hypotenuse of a primitive Pythagorean triple. We need a strategy for moving from primes to products of primes.

8.17 Exercise. *Check the following identity:*

$$(a^2 + b^2)(c^2 + d^2) = (ac + bd)^2 + (bc - ad)^2.$$

The preceding exercise tells us that the products of sums of two squares are themselves sums of two squares.

8.18 Theorem. *If an integer x can be written as the sum of two squares of natural numbers and an integer y can be written as the sum of two squares of natural numbers, then xy can be written as the sum of two squares of natural numbers.*

Let's try writing a few numbers as sums of squares of natural numbers.

8.19 Exercise. *For each of the following numbers, (i) determine the number's prime factorization and (ii) write the number as the sum of two squares of natural numbers.*

    1. 205

    2. 6409

3. 722

4. 11745

8.20 Question. *Which natural numbers can be written as the sum of two squares of natural numbers? State and prove the most general theorem possible about which natural numbers can be written as the sum of two squares of natural numbers, and prove it.*

We give the most general result next.

8.21 Theorem. *A natural number n can be written as a sum of two squares of natural numbers if and only if every prime congruent to 3 modulo 4 in the unique prime factorization of n occurs to an even power.*

Pythagorean triples revisited

We are now in a position to describe the possible values for the hypotenuse in a primitive Pythagorean triple.

8.22 Theorem. *If (a, b, c) is a primitive Pythagorean triple, then c is a product of primes each of which is congruent to 1 modulo 4.*

8.23 Theorem. *If the natural number c is a product of primes each of which is congruent to 1 modulo 4, then there exist integers a and b such that (a, b, c) is a primitive Pythagorean triple.*

Having satisfactorily analyzed the question of which squares are the sum of two smaller squares, it is natural to ask the analogous question for higher powers, and Pierre de Fermat did ask that question in what became known as Fermat's Last Theorem.

Fermat's Last Theorem

There are infinitely many Pythagorean triples of natural numbers (a, b, c) such that $a^2 + b^2 = c^2$. A natural question arises if we replace the exponent 2 with larger numbers. In other words, can we find triples of natural numbers (a, b, c) such that $a^3 + b^3 = c^3$ or $a^4 + b^4 = c^4$, or, in general, $a^n + b^n = c^n$ for $n \geq 3$? In 1637, Fermat claimed to be able to prove that no triple of natural numbers (a, b, c) exists that satisfies the equation $a^n + b^n = c^n$ for any natural number $n \geq 3$. During his lifetime, Fermat probably realized his "proof" was inadequate, but the question tantalized mathematicians for hundreds of years. Incremental progress was made. By 1992 it was known that the equations $a^n + b^n = c^n$ had no natural number solutions for $3 \leq n \leq 4,000,000$ (as well as many other special cases). But there are infinitely many possible exponents larger than $4,000,000$, so Fermat's Last Theorem was far from being resolved. But all the remaining exponents were taken care of by the groundbreaking work of Andrew Wiles, which took place some 350 years after Fermat first considered the question.

Theorem (Fermat's Last Theorem, proved by Andrew Wiles in 1994). *For natural numbers $n \geq 3$, there are no natural numbers x, y, z such that $x^n + y^n = z^n$.*

We probably won't find a proof of this theorem ourselves since it took many high-powered mathematicians 350 years to do so. Instead, let's look at one case of this theorem which can be proved using a strategy known as *Fermat's method of descent*. The method involves showing how a given solution in natural numbers can be used to produce a "smaller" natural number solution. That new solution would imply the existence of a yet smaller solution, and so on. Since any decreasing sequence of natural numbers must be finite in length, the method of descent implies that there could not be a solution to begin with. Let's see how this strategy can be used to prove the case of Fermat's Last Theorem when the exponent is 4.

In fact, notice that the following statement is a little stronger than what is called for in Fermat's Last Theorem since the z is squared rather than raised to the fourth power.

8.24 Theorem. *There are no natural numbers x, y, and z such that $x^4 + y^4 = z^2$.*

(Hint: Note that if there were a solution $x = a$, $y = b$, and $z = c$, then (a^2, b^2, c) would be a Pythagorean triple, which we could assume to be a primitive Pythagorean triple by removing common factors. Can you use the characterization of Pythagorean triples to find other natural numbers d, e, f such that $d^4 + e^4 = f^2$ where f is less than c? If you can do that, how can you complete your proof?)

8.25 Blank Paper Exercise. *After not looking at the material in this chapter for a day or two, take a blank piece of paper and outline the development of that material in as much detail as you can without referring to the text or to notes. Places where you get stuck or can't remember highlight areas that may call for further study.*

Who's Represented?

Representing numbers as the sum of two squares had immediate practical relevance to the description of Pythagorean triples. But it is also a problem that lends itself well to many different possible directions of generalization. For example,

1. Which numbers can be represented as the sum of three squares; sum of four squares; etc.?

2. Which numbers can be represented as the sum of two cubes; sum of two fourth powers; etc.?

Mathematicians have given much attention to all of these questions. This is another one of the many instances of simple sounding questions leading to deep and important mathematics.

Sums of squares

Albert Girard (1595–1632) appeared to know as early as 1625 which numbers could be written as the sum of two squares, although a proof due to Girard is lacking. Descartes proved in a 1638 letter to Mersenne that primes of the form $4n + 3$ could not be represented as a sum of two squares. Fermat stated in a letter to Blaise Pascal (1623–1662) in 1654 that he had a proof of the fact that primes of the form $4n + 1$ were always the sum of two squares. But a proof of Girard's complete (and correct) observation would have to wait for Euler, who gave a complete proof in two letters to Goldbach dated 1747 and 1749.

What about representing numbers as the sum of three squares? In a letter to Mersenne dated 1636, Fermat stated (again without proof!) that no integer of the form $8n + 7$ could be expressed as the sum of three squares. Mersenne communicated the claim to Descartes who provided a proof in 1638. The complete characterization is given here.

Theorem. *A natural number can be expressed as the sum of three squares of natural numbers if and only if it is not of the form $4^n(8k + 7)$ for non-negative integers n and k.*

The proof of this theorem is due in large part to Legendre, but a key step also requires Dirichlet's work on primes in arithmetic progressions.

What about sums of four squares? Fermat stated that he had a proof of the fact that every number is either a square or the sum of two, three, or four squares, although, as we now expect when dealing with Fermat, no proof was communicated. Building on the work of Fermat and Euler, it was Lagrange in 1770 who finally provided the proof of the following theorem.

Theorem (Four Squares Theorem). *Every natural number is the sum of at most four squares of natural numbers.*

A key identity needed for Lagrange's proof was due to Euler, who spent more than 40 years trying to establish the Four Squares Theorem. Euler established an amazing identity showing that the product of two numbers, each of which can be expressed as the sum of four squares, is also a sum of four squares, namely,

$$(a_1^2 + a_2^2 + a_3^2 + a_4^2)(b_1^2 + b_2^2 + b_3^2 + b_4^2)$$
$$= (a_1b_1 + a_2b_2 + a_3b_3 + a_4b_4)^2$$
$$+ (a_1b_2 - a_2b_1 + a_3b_4 - a_4b_3)^2$$
$$+ (a_1b_3 - a_2b_4 - a_3b_1 + a_4b_2)^2$$
$$+ (a_1b_4 + a_2b_3 - a_3b_2 - a_4b_1)^2.$$

Sums of cubes, taxicabs, and Fermat's Last Theorem

Euler, in 1770, provided us with a proof of the first case of Fermat's Last Theorem by establishing that no cube is the sum of two cubes. Of the numbers which *can* be expressed as the sum of two cubes, perhaps 1729 is the most famous.

Suffering from tuberculosis and lying in a hospital bed in London, the young Indian mathematician Ramanujan (1887–1920) was paid a visit by his friend and mentor G. H. Hardy (1877–1947). Hardy remarked that he had arrived in a taxicab numbered 1729, which he considered a rather dull number. Ramanujan responded that 1729 is not dull at all. It is, in fact, the smallest number that can be expressed as the sum of two cubes in two essentially distinct ways,

$$1729 = 1^3 + 12^3 = 9^3 + 10^3.$$

Said another way, there are (at least) four distinct integer points, namely $(1, 12)$, $(12, 1)$, $(9, 10)$, and $(10, 9)$, on the cubic plane curve

$$x^3 + y^3 = 1729.$$

Taking statements about numbers and transforming them into statements about points on curves (or surfaces, etc.) is now a fairly common practice in the field of *arithmetical geometry*. For example, in studying whether the number m is expressible as a sum of two cubes, the corresponding plane curve is given by

$$x^3 + y^3 = m.$$

This is another example of what is known as an *elliptic curve*. While naturally arising when looking at the problem of expressing a number as the sum of two cubes, elliptic curves have also played a much more central role in the modern development of number theory. They are the central objects under study in Andrew Wiles' proof of Fermat's Last Theorem.

In 1990 it was known that if (a, b, c) were a triple of natural numbers satisfying an equation of the form

$$a^p + b^p = c^p,$$

where p is a prime greater than 2 (i.e., if the triple (a, b, c) provided a counterexample to Fermat's Last Theorem), then the curve

$$y^2 = x(x - a^p)(x + b^p)$$

would be an elliptic curve with some very strange properties. The precise statement is that the curve would be *semistable* but not *modular*, although the exact meanings of these words is beyond the scope of this book. Such a curve was believed not to exist. More precisely, it was believed by many (and was the content of the Shimura-Taniyama Conjecture) that *all* elliptic curves were modular. This conjecture is now known to be true. The first major contribution to the proof of the Shimura-Taniyama Conjecture was due to Wiles with the help of his student Richard Taylor. Wiles and Taylor proved in 1994 that *all semistable elliptic curves are modular*, once and for all confirming the truth of Fermat's Last Theorem.

9

Rationals Close to Irrationals and the Pell Equation

Diophantine Approximation and Pell Equations

Linear Diophantine equations were considered and solved in Chapter 1. In the previous chapter we asked which natural numbers could be written as the sum of two squares. That is, we sought solutions to the quadratic Diophantine equation $x^2 + y^2 = n$ which in turn gave us a complete description of the natural numbers that could occur as the hypotenuse in a primitive Pythagorean triple. In this chapter we consider one additional family of quadratic Diophantine equations called Pell equations. A Pell equation is any equation of the form $x^2 - Ny^2 = 1$ where N is any natural number. These equations have surprising connections to at least two different issues. One is a famous Bovine Problem about herds of cows and bulls whose sizes are related in various ways. This story problem was framed by Archimedes (287–212 B.C.) in the third century B.C. and was not completely solved until 1965. The minimum number of cattle that would satisfy the conditions of Archimedes' problem is vastly greater than the number of atoms in the universe, so you may not encounter all of them during the running of the bulls.

On a less frivolous note, the so-called Pell equations are also connected with the subject of *Diophantine approximation*; namely, the study of rational number approximations to irrational quantities. Of course, every irrational number can be arbitrarily closely approximated by rational numbers by just truncating the decimal representation of the irrational number, but here we consider the question of finding rational approximations where the size of the denominator of the approximating fraction is small relative to how close the approximation is. One challenge is to clarify the questions about rational approximations. Then we will find that the Pell equations, $x^2 - Ny^2 = 1$, help us analyze good rational approximations of certain irrational numbers.

Unfortunately, the name of the Pell equations is a misnomer. Mathematician John Pell (1611–1685) had little if anything to do with the study of the equations which now bear his name. In a published paper Euler mistakenly attributed what is believed to be the work of William Brouncker (1620–1684) to Pell, and the name has stuck. So there are at least two roads to mathematical immortality—prove something great or have a famous person think you proved it.

A plunge into rational approximation

Irrational numbers can sometimes pose a problem when it comes to practical computation. In practice, we always have to rely on rational approximations when irrationals are involved. We have all used close rational approximations in order to simplify and expedite solutions to problems that involve irrational numbers. For example, 1.414 is a convenient approximation for $\sqrt{2}$; and 3.14 or $\frac{22}{7}$ are often used as approximations for π. In fact, wise political minds have not overlooked the advantages of rational approximations to π. At times politicians have considered cutting the Gordian Knot by legislating π to equal a convenient rational value. In 1897, the Indiana Legislature considered and nearly accomplished the passage of such legislation; however, after being recommended for passage by the Committee on Education and passed by the House, a mathematician gave some advice that derailed this progressive legislation and the bill floundered in the Senate. Too bad.

Let's begin our investigation into rational approximations of irrational numbers by observing that it is an easy matter to approximate irrational numbers by fractions $\frac{a}{b}$ that lie within $\frac{1}{2b}$ of the irrational. Recall that the quantity $|x - y|$ measures the distance between the numbers x and y.

9.1 Theorem. *Let α be an irrational number and let b be a natural number. Then there exists an integer a such that*

$$|\alpha - \frac{a}{b}| \leq \frac{1}{2b}.$$

So a harder challenge of rational approximation is to find fractions $\frac{a}{b}$ that lie within a smaller distance of the target irrational, for example, within $\frac{1}{b^2}$. One technique for finding such approximations involves noticing that in any large collection of real numbers, some pair of them must have a difference that is close to being an integer in value. We begin by considering multiples of $\sqrt{2}$ and asking you to find a way to produce a good rational approximation to $\sqrt{2}$.

9.2 Exercise. *Among the first eleven multiples of* $\sqrt{2}$,

$$0\sqrt{2},\ \sqrt{2},\ 2\sqrt{2},\ 3\sqrt{2},\ \ldots,\ 10\sqrt{2},$$

find the two whose difference is closest to a positive integer. Feel free to use a calculator. Use those two multiples to find a good rational approximation for $\sqrt{2}$. *By good, we mean that you find integers a and b such that*

$$\left|\frac{a}{b} - \sqrt{2}\right| \leq \frac{1}{b^2}.$$

The technique of using a list of integer multiples to obtain good approximations to an irrational number is a valuable strategy to understand well. So after doing the previous exercise, think carefully about your method to see how generally the method can be applied and how each step was involved in the solution. To understand the method, do it once again for $\sqrt{7}$.

9.3 Exercise. *Repeat the previous exercise for* $\sqrt{7}$ *using the first 13 multiples of* $\sqrt{7}$.

Before we move along any further, was it important in the previous two exercises that the irrational being approximated was a square root?

9.4 Exercise. *Repeat the previous exercise for* π, *using the first 15 multiples of* π.

Now take some time to think through what you have done and why it works. By considering the following questions you are exploring how the preceding specific examples can be extended to apply to more general cases.

9.5 Question. *Let* α *be an irrational number.*

1. *Imagine making a list of the first 11 multiples of* α. *Can you predict how close to an integer the nearest difference between two of those numbers must be?*

2. *Now imagine making a list of 11 multiples of* α, *but not the first 11. Can you still predict how close to an integer the nearest difference between two of those numbers must be?*

3. *Now imagine making a list of* 50 *multiples of* α, *rather than just* 11. *Can you predict how close to an integer the nearest difference between two of those numbers must be?*

4. *What is the general relationship between how many multiples of* α *we consider and how well we can rationally approximate* α *using our multiples?*

The next three theorems formalize what you may have discovered in the preceding group of questions.

9.6 Theorem. *Let* K *be a positive integer. Then, among any* K *real numbers, there is a pair of them whose difference is within* $1/K$ *of being an integer.*

When we take our collection of real numbers to be multiples of an irrational number, then we can find good rational approximations for the irrational number. Remember how multiples of an irrational could lead to rational approximations of the irrational by finding multiples whose difference is close to an integer.

9.7 Theorem. *Let* α *be a positive irrational number and* K *be a positive integer. Then there exist positive integers* a, b, *and* c *with* $0 \leq a < b \leq K$ *and* $0 \leq c \leq K\alpha$ *such that*

$$\left| \frac{c}{b-a} - \alpha \right| \leq \frac{1}{(b-a)^2}.$$

Theorem 9.6 told us that increasingly large collections of real numbers contain pairs whose differences get increasingly close to being an integer. Now you will need to understand your proof of the above theorem sufficiently well so that you can figure out how to make $(b-a)$ arbitrarily large. You might consider the fact that for an irrational number α, any fixed, finite collection of multiples of α will have every difference of every pair of those multiples differing from being an integer by at least some specific non-zero amount. So taking a yet bigger collection of multiples will give you a pair whose difference is even closer to being an integer. That observation might help to generalize your technique to prove Dirichlet's Rational Approximation Theorem.

9.8 Theorem (Dirichlet's Rational Approximation Theorem, Version I). *Let* α *be any real number. Then there exist infinitely many rational numbers* $\frac{a}{b}$

satisfying

$$\left|\frac{a}{b} - \alpha\right| \leq \frac{1}{b^2}.$$

It is often useful to put the same result in different forms, because the different forms might help us to see a connection with some other work. In this case, the following alternative form of Dirichlet's Rational Approximation Theorem takes the first step toward making the connection between rational approximation and Pell's equation.

Theorem (Dirichlet's Rational Approximation Theorem, Version II). *Let α be any real number. Then there exist infinitely many integers a and b satisfying*

$$|a - b\alpha| \leq \frac{1}{b}.$$

Before going further, let's confirm that these two versions of Dirichlet's Rational Approximation Theorem actually are equivalent.

9.9 Theorem. *Show that Versions I and II of Dirichlet's Rational Approximation Theorem can be deduced from one another.*

If we consider the special case where α is the square root of a natural number, we get a form of Dirichlet's Rational Approximation Theorem that looks even more like Pell's Equation.

Theorem (Dirichlet's Rational Approximation Theorem, Version III). *Let N be a positive integer that is not a square. Then there exist infinitely many positive integers a and b satisfying*

$$\left|a - b\sqrt{N}\right| \leq \frac{1}{b}.$$

The connection between Pell equations and rational approximations to irrational numbers that are square roots of natural numbers is not hard to make.

9.10 Exercise. *Show that if N is a natural number which is not a square and x = a and y = b is a positive integer solution to the Pell equation $x^2 - Ny^2 = 1$, then $\frac{a}{b}$ gives a good rational approximation to $\sqrt{N}$.*

The next theorem clarifies that by a "good" rational approximation we mean the same thing that occurs in Dirichlet's Theorem Version I.

9.11 Theorem. *Let N be a positive integer that is not a square. If $x = a$ and $y = b$ is a solution in positive integers to $x^2 - Ny^2 = 1$, then*

$$\left| \frac{a}{b} - \sqrt{N} \right| < \frac{1}{b^2}.$$

So we see that solutions in positive integers to the Pell equation $x^2 - Ny^2 = 1$ give rise to good approximations to the irrational number $\sqrt{N}$. So our challenge now is to analyze the Pell equation and see whether we can find solutions. We'll start by disposing of trivial cases so that we can focus on the ones that count.

Out with the trivial

In Chapter 1 we considered the family of linear Diophantine equations

$$ax + by = c.$$

Certain values of the parameters a, b, and c lead to Diophantine equations with no hope of having solutions. For example, the equation $6x + 3y = 17$ will not have any integer solutions because the left-hand side will always be divisible by 3, and the right-hand side will never be divisible by 3.

When working with a parameterized family of equations, it is worthwhile making an effort to recognize whether certain values of the parameters will lead to obvious conclusions or whether there are some trivial solutions that are not of interest. Let's try this with the Pell equations $x^2 - Ny^2 = 1$, which have the single parameter, the natural number N.

9.12 Question. *For every natural number N, there are some trivial values of x and y that satisfy the Pell equation $x^2 - Ny^2 = 1$. What are those trivial solutions?*

Let's pin that down by making the following definitions of trivial and non-trivial solutions.

Definition. Let N be a natural number. The *trivial solutions* to the Diophantine equation $x^2 - Ny^2 = 1$ are $x = 1$, $y = 0$ and $x = -1$, $y = 0$. All other integer solutions are *non-trivial*.

9.13 Question. *For what values of the natural number N can you easily show that there are no non-trivial solutions to the Pell equation $x^2 - Ny^2 = 1$?*

We record your observation in the following theorem.

9.14 Theorem. *If the natural number N is a perfect square, then the Pell equation*

$$x^2 - Ny^2 = 1$$

has no non-trivial integer solutions.

After all this talk about trivial solutions, let's at least confirm that in some cases non-trivial solutions do exist.

9.15 Exercise. *Find, by trial and error, at least two non-trivial solutions to each of the Pell equations $x^2 - 2y^2 = 1$ and $x^2 - 3y^2 = 1$.*

Bolstered by the existence of solutions for $N = 2$ and $N = 3$, our focus from this point forward will be on finding non-trivial solutions to the Pell equations $x^2 - Ny^2 = 1$ where N is a natural number that is not a perfect square.

New solutions from old

For a positive integer N that is not a perfect square, the non-trivial solutions to $x^2 - Ny^2 = 1$ come to us in natural groups of four since the square of a negative number is positive.

9.16 Question. *To know all the integer solutions to a Pell equation, why does it suffice to know just the positive integer solutions?*

One solution to a Pell equation gives rise to related ones by taking negatives, but there are other ways to take some solutions and combine them to create other solutions. Since 1 times 1 equals 1, multiplication of solutions also gives a new solution. Here is what we mean.

9.17 Theorem. *Suppose N is a natural number and the Pell equation $x^2 - Ny^2 = 1$ has two solutions, namely, $a^2 - Nb^2 = 1$ and $c^2 - Nd^2 = 1$ for some integers a, b, c, and d. Then $x = ac + Nbd$ and $y = ad + bc$ is also an integer solution to the Pell equation $x^2 - Ny^2 = 1$. That is,*

$$(ac + Nbd)^2 - N(ad + bc)^2 = 1.$$

So we can generate new solutions to the Pell equation from old solutions. But the question remains: For which positive integers N (which are not squares) does $x^2 - Ny^2 = 1$ have non-trivial solutions? To fully answer this question we return to the world of rational approximation.

Securing the elusive solution

We observed earlier that non-trivial solutions to the Pell equation $x^2 - Ny^2 = 1$ give rise to good approximations of $\sqrt{N}$. Now we look at the connection between good rational approximations of $\sqrt{N}$ and Pell-like equations in the opposite way. That is, starting with a "good" rational approximation $\frac{x}{y}$ of $\sqrt{N}$, let's investigate $x^2 - Ny^2$. Recall Version II of Dirichlet's Rational Approximation Theorem. That version described the closeness of the rational approximation of the fraction $\frac{x}{y}$ to $\sqrt{N}$ by stating that $|x - y\sqrt{N}| < \frac{1}{y}$. That concept of a good rational approximation is used as the hypothesis in the following theorem.

9.18 Theorem. *Let N be a natural number and suppose that x and y are positive integers satisfying $|x - y\sqrt{N}| < \frac{1}{y}$. Then*

$$x + y\sqrt{N} < 3y\sqrt{N}.$$

A tiny bit of algebra gets us back to a Pell-like expression.

9.19 Theorem. *Let N be a natural number and suppose that x and y are positive integers satisfying $|x - y\sqrt{N}| < \frac{1}{y}$. Then*

$$|x^2 - Ny^2| < 3\sqrt{N}.$$

Notice that the preceding theorem tells us that any good rational approximation of $\sqrt{N}$ gives rise to a Pell-like expression, $|x^2 - Ny^2|$, which is an integer with a fixed bound. We want to find solutions to the Pell equation $x^2 - Ny^2 = 1$; however, let's take what we can get at this point, namely, solutions to a Pell-like equation where the right side is some integer possibly different from 1.

9.20 Theorem. *There exists a non-zero integer K such that the equation*

$$x^2 - Ny^2 = K$$

has infinitely many solutions in positive integers.

Now we have infinitely many positive integer solutions to a Pell-like equation,

$$x^2 - Ny^2 = K.$$

In the next few theorems we investigate how to use these to obtain a non-trivial solution to

$$x^2 - Ny^2 = 1.$$

9.21 Lemma. *Let n be a natural number and suppose that (x_i, y_i), $i = 1, 2, 3, \ldots$ are infinitely many ordered pairs of integers. Then there exist distinct natural numbers j and k such that*

$$x_j \equiv x_k \pmod{n} \quad and \quad y_j \equiv y_k \pmod{n}.$$

9.22 Lemma. *Let N be a natural number and K be a non-zero integer and let (x_j, y_j) and (x_k, y_k) be two distinct integer solutions to $x^2 - Ny^2 = K$ satisfying*

$$x_j \equiv x_k \pmod{|K|} \quad and \quad y_j \equiv y_k \pmod{|K|}.$$

Then

$$x = \frac{x_j x_k - y_j y_k N}{K} \quad and \quad y = \frac{x_j y_k - x_k y_j}{K}$$

are integers satisfying $x^2 - Ny^2 = 1$.

What you have now proved is that the Pell equation $x^2 - Ny^2 = 1$ has non-trivial solutions for every possible case, namely for any natural number N that is not a perfect square.

9.23 Theorem. *If N is a positive integer that is not a square, then the Pell equation $x^2 - Ny^2 = 1$ has a non-trivial solution in positive integers.*

An excellent way to understand a proof is to follow the steps of the proof for some particular examples. That is what we ask you to do in the next exercise.

9.24 Exercise. *Follow the steps of the preceding theorems to find several solutions to the Pell equations $x^2 - 5y^2 = 1$ and $x^2 - 6y^2 = 1$ and then give some good rational approximations to $\sqrt{5}$ and $\sqrt{6}$.*

The structure of the solutions to the Pell equations

We have now proved that the Pell equations have solutions, but in fact those solutions have a satisfying kind of structure to them, which we will explore in this section. This structure arises from our inability to resist factoring when we have the chance.

The left sides of the Pell equations $x^2 - Ny^2 = 1$ look very much like the difference of two squares. It is difficult to see a difference of two squares without succumbing to the urge to factor. Giving in to that temptation pays off in this case. Of course, there is one unpleasant part of that factoring, namely, when N is not a perfect square, the factors involve an irrational number, $\sqrt{N}$. Never mind, let's factor anyway.

$$x^2 - Ny^2 = 1,$$

$$\left(x + y\sqrt{N}\right)\left(x - y\sqrt{N}\right) = 1.$$

Definition. Let N be a natural number. We say that a real number $\alpha = r + s\sqrt{N}$, with r and s integers, *gives a solution* to the Pell equation $x^2 - Ny^2 = 1$ if $r^2 - Ns^2 = 1$.

The next several theorems work out the algebraic structure of the real numbers that give integer solutions to a given Pell equation.

9.25 Theorem. *Let N be a natural number and r_1, r_2, s_1, and s_2 be integers. If $\alpha = r_1 + s_1\sqrt{N}$ and $\beta = r_2 + s_2\sqrt{N}$ both give solutions to the Pell equation $x^2 - Ny^2 = 1$, then so does $\alpha\beta$.*

9.26 Theorem. *Let N be a natural number and r and s integers. If $\alpha = r + s\sqrt{N}$ gives a solution to $x^2 - Ny^2 = 1$, then so does $1/\alpha$.*

Note: Abstract algebra is a study of algebraic structures and relationships. When you study abstract algebra, one of the first structures you will encounter is a *group*. We won't define the idea of a group here, but the previous two theorems tell us that the set of real numbers of the form $r + s\sqrt{N}$, with r and s integers, which give solutions to the Pell equation $x^2 - Ny^2 = 1$, form a group with respect to the operation of multiplication.

9.27 Corollary. *Let N be a natural number and r and s integers. If $\alpha = r + s\sqrt{N}$ gives a solution to $x^2 - Ny^2 = 1$, then so does α^k for any integer k.*

9.28 Exercise. *Let N be a natural number and r and s integers. Show that if $r + s\sqrt{N}$ gives a solution to $x^2 - Ny^2 = 1$, then so do each of*

$$r - s\sqrt{N}, \quad -r + s\sqrt{N}, \quad \text{and} \quad -r - s\sqrt{N}.$$

9.29 Theorem. *Let N be a positive integer that is not a square. Let A be the set of all real numbers of the form $r + s\sqrt{N}$, with r and s positive* *integers, that give solutions to $x^2 - Ny^2 = 1$. Then*

1. *there is a smallest element α in A,*

2. *the real numbers α^k, $k = 1, 2, \ldots$ give all positive integer solutions to $x^2 - Ny^2 = 1$.*

(Hint: For part (1), try showing that the numbers in question are ordered by r. Then use the Well-Ordering Axiom.)

Let's reflect on what we have shown so far. If the natural number N is a perfect square, then the Pell equation $x^2 - Ny^2 = 1$ has only trivial solutions. In all other cases, it suffices to focus on just the positive integer solutions. In these cases, Theorem 9.23 tells us that there is a non-trivial solution and Theorem 9.29 suggests that in a sense there is a "smallest" solution in positive integers, which generates all of the infinitely many other positive integer solutions. So our investigation of the Pell equations has revealed a satisfying mathematical structure.

9.30 Blank Paper Exercise. *After not looking at the material in this chapter for a day or two, take a blank piece of paper and outline the development of that material in as much detail as you can without referring to the text or to notes. Places where you get stuck or can't remember highlight areas that may call for further study.*

Bovine Math

Pell equations are not merely mathematical amusements. They also arise in ranching by the gods. The following is an English translation, due to Ivor Thomas, of the *problema bovinum* attributed to Archimedes. It is written in the form of a challenge, and considers the number of four different types of cattle belonging to the herd of the sun god Helios.

If thou art diligent and wise, O stranger, compute the number of cattle of the Sun, who once upon a time grazed on the fields of the Thrinacian isle of Sicily, divided into four herds of different colours, one milk white, another a glossy black, a third yellow and the last dappled. In each herd were bulls, mighty in number according to these proportions: Understand, stranger, that the white bulls were equal to a half and a third of the black together with the whole of the yellow, while the black were equal to the fourth part of the dappled and a fifth, together with, once more, the whole of the yellow. Observe further that the remaining bulls, the dappled, were equal to a sixth part of the white and a seventh, together with all of the yellow. These were the proportions of the cows: The white were precisely equal to the third part and a fourth of the whole herd of the black; while the black were equal to the fourth part once more of the dappled and with it a fifth part, when all, including the bulls, went to pasture together. Now the dappled in four parts were equal in number to a fifth part and a sixth of the yellow herd. Finally the yellow were in number equal to a sixth part and a seventh of the white herd. If thou canst accurately tell, O stranger, the number of cattle of the Sun, giving separately the number of well-fed bulls and again the number of females according to each colour, thou wouldst not be called unskilled or ignorant of numbers, but not yet shalt thou be numbered among the wise.

But come, understand also all these conditions regarding the cattle of the Sun. When the white bulls mingled their number with the black, they stood firm, equal in depth and breadth, and the plains of Thrinacia, stretching far in all ways, were filled with their multitude. Again, when the yellow and the dappled bulls were gathered into one herd they stood in such a manner that their number, beginning from one, grew slowly greater till it completed a triangular figure, there being no bulls of other colours in their midst nor none of them lacking. If thou art able, O stranger, to find out all these things and gather them together in your mind, giving all the relations, thou shalt depart crowned with glory and knowing that thou hast been adjudged perfect in this species of wisdom.

How can we hope to be crowned with glory? Obviously, we must get our cows and bulls in a row, steer clear of mooving mooers, and solve this bully conundrum.

The first paragraph translates mathematically into a system of seven linear equations in 8 unknowns (the four types of bulls: W, B, Y, D, and the four types of cows: w, b, y, d). There is a 1-parameter family of solutions given by

$$W = 10366482k \qquad w = 7206360k$$
$$B = 7460514k \qquad b = 4893246k$$
$$Y = 4149387k \qquad y = 5439213k$$
$$D = 7358060k \qquad d = 3515820k$$

The second paragraph imposes two additional conditions: the sum of the white bulls and the black bulls should be a square, and the sum of the yellow bulls and the dappled bulls should be a *triangular number*, that is, a number of the form $1 + 2 + \cdots + m = m(m + 1)/2$. These constraints tell us that

$$W + B = 10366482k + 7460514k = 17826996k = n^2 \qquad (1)$$

for some integer n, and

$$Y + D = 4149387k + 7358060k = 11507447k = \frac{m(m + 1)}{2} \qquad (2)$$

for some integer m. The factorization $17826996 = 2^2 \cdot 3 \cdot 11 \cdot 29 \cdot 4657$ tells us that the value of k in equation (1) must be of the form

$$k = 3 \cdot 11 \cdot 29 \cdot 4657 \cdot y^2 = 4456749y^2$$

for some integer y. Combining this with the equation (2) gives

$$11507447 \cdot 4456749y^2 = \frac{m(m + 1)}{2},$$

or

$$51285802909803y^2 = \frac{m(m + 1)}{2}. \qquad (3)$$

Completing the square on the right-hand side of equation (3) we obtain

$$\frac{m(m + 1)}{2} = \frac{(m + 1/2)^2 - 1/4}{2} = \frac{1}{8}\left((2m + 1)^2 - 1\right).$$

So, by multiplying equation (3) by 8, and making the substitution $x = 2m + 1$ we obtain

$$8 \cdot 51285802909803y^2 = x^2 - 1,$$

or

$$x^2 - 410286423278424y^2 = 1,$$

a Pell equation!

Our translation of the cattle problem into a Pell equation is unlikely to have been employed during Archimedes' time. And even more unlikely is it that he, or any of his contemporaries, produced a solution, even though we now know that in fact there are infinitely many. The first known complete solution, aided by computers, was given in 1965 by H. C. Williams, R.

A. German, and C. R. Zarnke. The smallest sized herd satisfying all the conditions is so vast that to write down the number of cattle we would need to use 206545 digits! That's a lot of bulls. To put that number in perspective, the number of atoms in the universe is estimated to be described with a number with a mere 80 digits.

Archimedes was not the only mathematician to issue challenges. Fermat was known to challenge his contemporaries as well. In 1657 he sent letters asking William Brouncker and John Wallis (1616–1703) to find integer solutions to the equations $x^2 - 151y^2 = 1$ and $x^2 - 313y^2 = 1$. Both stepped up to the challenge and gave integer solutions in reply.

But it is in early Indian mathematics that we find the first systematic study of Pell equations. Brahmagupta was aware of how to generate new solutions from old in much the same manner as we explored in Theorem 9.17, and both Brahmagupta and Bhaskara (1114–1185) discovered methods for turning solutions of $x^2 - Ny^2 = K$ (for small K) into solutions to $x^2 - Ny^2 = 1$. So Pell equations have spanned the ages, spanned the globe, and have even amused the sun god.

10

The Search for Primes

Primality Testing

Determining whether or not a large number is prime has practical importance in cryptography as seen in Chapter 5. If a number is relatively small, we might try simple trial division up to its square root (see Theorem 2.3). If we find no divisor, we have a prime. But trial division quickly becomes an overwhelming burden. Trial division on a large number, say with 100's of digits, would take today's fastest computers longer than the entire history of the universe since the Big Bang 13.6 billion years ago. That is too long to wait. So trial division is not a *fast* algorithm for determining primality.

Is it prime?

In this section we look at the notion of a *primality test*. We also examine just exactly what mathematicians mean when describing an algorithm as "fast."

To be precise, by a *primality test* we mean a theorem of the form

A natural number n is prime if and only if _____.

where the blank would be filled in by some testable condition on n. For example

Theorem. *A natural number n is prime if and only if for all primes $p \leq \sqrt{n}$, p does not divide n.*

Although this theorem provides a primality test, it does little to help our agent in the field set up a secure RSA public key code system. It is completely impractical for identifying, say, 200 digit primes. In Chapter 4 we find the following primality test.

Theorem (Wilson's Theorem and Converse). *A natural number n is prime if and only if $(n-1)! \equiv -1 \pmod{n}$.*

Unfortunately, there are no general shortcuts for computing $(n-1)!$ $\pmod{n}$, and as n begins to grow, even our fastest computers become overwhelmed with the computation.

Mathematicians measure the speed or complexity of a primality testing algorithm as a function of the number of digits in the number to be tested.

10.1 Exercise. *If n is a d-digit number, explain why the trial division primality test requires roughly $10^{d/2}$ trials.*

10.2 Exercise. *If n is a d-digit number, explain why the Wilson's Theorem primality test requires roughly 10^d multiplications.*

These two algorithms are said to run in *exponential time* since the required number of steps is an exponential function in the number of digits in the number to be tested. Exponential time algorithms are considered slow, and quickly become impractical for modern computers to carry out. A faster class of algorithms are those which run in *polynomial time*, that is, those for which the number of required steps is a polynomial function in the number of digits. Just how much of a difference does polynomial time versus exponential time make?

10.3 Question. *Suppose that Algorithm A requires d^2 steps and Algorithm B requires 2^d steps, where d is the number of digits in the number to be tested. Suppose our computer can carry out one million steps per second. How long would it take for our computer to carry out each algorithm when the number to be tested has 200 digits?*

Fermat's Little Theorem and probable primes

Both primality tests given in the preceding section are impractical for identifying really large primes. On the other hand, computing powers modulo n is an operation we have seen to be fast even for large numbers. In fact, in Chapter 3 you discovered that the computation of $a^r \pmod{n}$ requires roughly $\log_2 r$ multiplications.

10.4 Exercise. *Show that the algorithm described in Question 3.6 for computing a^r (mod n) is a polynomial time algorithm in the number of digits in r.*

In the next series of problems you will explore the use of this operation as a means of testing for primality by starting with a familiar theorem.

Theorem (Fermat's Little Theorem). *Let p be a prime. Then for all natural numbers a less than p, $a^{p-1} \equiv 1$ (mod p).*

Fermat's Little Theorem can be useful for showing certain numbers are composite.

10.5 Exercise. *State the contrapositive of Fermat's Little Theorem.*

10.6 Exercise. *Use Fermat's Little Theorem to show that $n = 737$ is composite.*

Unfortunately, the statement of Fermat's Little Theorem lacks the logical connective "if and only if" that we desire for a primality test. This raises the question of whether the converse to Fermat's Little Theorem is true.

10.7 Question. *State the converse to Fermat's Little Theorem. Do you think the converse to Fermat's Little Theorem is true?*

10.8 Theorem. *Let n be a natural number greater than 1. Then n is prime if and only if $a^{n-1} \equiv 1$ (mod n) for all natural numbers a less than n.*

10.9 Question. *Does the previous theorem give a polynomial or exponential time primality test?*

Inventing polynomial time primality tests is quite a challenge. One way to salvage some good from Fermat's Little Theorem is to weaken our demand of certainty. What if instead we look for a *probable prime test*, by which we mean a statement of the form

       If _____, then n is very likely to be prime.

where the blank would be filled in by some testable condition on n.

10.10 Exercise. *Compute 2^{n-1} (mod n) for all odd numbers n less than 100. If you have access to a computer, and some computing software, keep going. Test any conjectures you make along the way. State a probable prime test based on your observations.*

The evidence you collected hopefully suggests the following probable prime test for natural numbers n bigger then 2.

$$\text{If } \begin{cases} 2^{n-1} \not\equiv 1 \pmod{n}, & \text{then } n \text{ is composite,} \\ 2^{n-1} \equiv 1 \pmod{n}, & \text{then } n \text{ is very likely prime.} \end{cases}$$

We cannot remove the words "very likely" in this probable prime test because there are composite numbers n for which $2^{n-1} \equiv 1 \pmod{n}$. The first composite that fools our probable prime test is $341 = 11 \cdot 31$. Composite numbers n such that $2^{n-1} \equiv 1 \pmod{n}$ are sometimes called *Poulet numbers*. There are infinitely many, but they are so rare that for practical purposes, most people feel completely comfortable using our probable prime test to identify large primes.

For example, if n is a randomly chosen 13 digit odd number and $2^{n-1} \equiv 1 \pmod{n}$, then there is a 99.9999996% chance that n is prime, because there are 308457624821 13 digit primes and 132640 13 digit Poulet numbers. Would you feel safe with those odds? At a cost of guaranteed certainty, we now have a polynomial time probable prime test!

AKS primality

There are many polynomial time probable prime tests, but it was not known until the summer of 2002 whether or not a polynomial time primality test could exist. That summer an Indian scientist and two of his undergraduate students made public their discovery of a deterministic polynomial time primality test. Manindra Agrawal and his students Neeraj Kayal and Nitin Saxena would eventually win the Gödel prize in computer science for their work.

The test, now know as the AKS primality test, is based on the following theorem.

10.11 Theorem. *Let a and n be relatively prime natural numbers. Then n is prime if and only if $(x + a)^n \equiv x^n + a \pmod{n}$ for every integer x.*

This theorem alone constitutes a primality test, but a slow one at that. The problem lies in the fact that there are n different coefficients to compute in $(x + a)^n \pmod{n}$. Part of what Agrawal, Kayal, and Saxena were able to figure out is how to reduce the degree of the polynomials that need to be checked.

The polynomial time deterministic AKS primality test may be beyond the scope of this book, but please do not assume that it is beyond the scope of your abilities. With a little bit of abstract algebra and the number theory you have learned so far you'll be more than prepared to tackle the AKS primality test for yourself.

10.12 Blank Paper Exercise. *After not looking at the material in this chapter for a day or two, take a blank piece of paper and outline the development of that material in as much detail as you can without referring to the text or to notes. Places where you get stuck or can't remember highlight areas that may call for further study.*

Record Primes

A list of the largest known primes will show that they all share the following property: each prime is either 1 more or 1 less than an easily factored number. In September, 2006, the largest known prime was

$$2^{32582657} - 1,$$

which is a Mersenne prime with over 9.8 million digits. Clearly it is 1 less than a very easily factored number. In fact, the six largest known primes are Mersenne primes (again, as of September 2006), and the seventh largest is

$$27653 \cdot 2^{9167433} + 1,$$

which is 1 more than an easily factored number (27653 is prime). This fact is not just coincidence. When n is a natural number of a certain special form, much more efficient primality tests are available for determining the nature of n. In this section we present some of these wonderful theorems that have helped people discover some of the largest known primes.

The late nineteenth century witnessed tremendous progress in the mathematics of primality testing. Edouard Lucas (1842–1891) was one of the thinkers who concerned themselves with such matters. The n-th Fermat number is given by $F_n = 2^{2^n} + 1$. Fermat had determined that F_1, F_2, F_3, and F_4 are each prime and conjectured that every Fermat number was prime (although he didn't call them Fermat numbers). In 1732 Euler proved that Fermat's conjecture was false by showing that $F_5 = 4294967297$ is divisible by 641. But the nature of F_6 remained unresolved until Lucas developed a primality test for Fermat numbers that proved that F_6 is also composite.

Father Theophile Pepin (1826–1905), a contemporary of Lucas, published another primality test for Fermat numbers in 1877 which still bears his name.

Theorem (Pepin's Test). *Let F_n denote the n-th Fermat number. Then F_n is prime if and only if*

$$3^{(F_n-1)/2} \equiv -1 \pmod{F_n}.$$

In Pepin's original theorem the condition appears as $5^{(F_n-1)/2} \equiv -1$ (mod F_n). It was another contemporary, Francois Proth (1852–1879), who pointed out that 3 would work as well as 5. Proth contributed primality tests of his own as well, which have been implemented today (see Yves Gallot's Proth.exe) and are responsible for finding some of the currently largest known primes (at least those that are not Mersenne primes). Proth's 1878 test is as follows.

Theorem (Proth's Test). *Let n and k be natural numbers, and let $N = k \cdot 2^n + 1$ with $2^n > k$. If there is an integer a such that*

$$a^{(N-1)/2} \equiv -1 \pmod{N},$$

then N is prime.

So what about the record-holding Mersenne primes? In 1930 D. H. Lehmer (1905–1991) completed a dissertation at Brown University titled *An Extended Theory of Lucas' Functions*. In it, we find the following test, which is responsible for identifying today's largest known primes. The form of this theorem is similar to that of Lucas' earlier primality tests for Fermat numbers.

Theorem (Lucas-Lehmer Test). *Let $M_n = 2^n - 1$ denote the n-th Mersenne number, and define the sequence $\{S_i\}$ by*

$$S_0 = 4, \quad S_{i+1} = S_i^2 - 2.$$

Then M_n is prime if and only if $S_{n-2} \equiv 0 \pmod{M_n}$.

Since there are infinitely many primes, the quest for ever larger primes is an endless pursuit. The current strategies for finding such primes involve having many computers, contributed by volunteers around the world, work in concert to find new, huge primes. Number theory has had unexpected applications to cryptography, as we saw in Chapter 5. Perhaps an unexpected consequence of the search for large primes will be the development of previously unimagined strategies for global cooperation.

A

Mathematical Induction: The Domino Effect

The Infinitude Of Facts

Many mathematical theorems are really infinitely many little theorems all packaged into one statement. For example, we learn the following theorem in calculus: *Every polynomial function is continuous.* If you were lucky enough to also see a proof of this theorem, you would know that we did not separately consider *every* polynomial. If we did, you would still be sitting in that calculus class. One of the great strengths of mathematical reasoning and logic is the ability to prove an infinite number of facts in a finite amount of space and time.

Gauss' formula

Carl Friedrich Gauss was a famous mathematician of the early 19th century. A story about his boyhood has made its way into mathematical folklore. As the story goes, an elementary school teacher of Gauss wanted to keep his students busy while he graded papers. To this end, he asked his students to add up the first one hundred numbers, thinking this task would keep them quiet for a long time. To the dismay of the teacher, Gauss quickly discovered a shortcut to replace the tedious addition problem and came up with the answer after only a few short moments. As a cultural aside, historians feel that this story is probably false, and some feel that it promotes the false myth that mathematics is a subject only for the rare genius rather than for everybody. Regardless of the historical or political status of the story, the technique for adding the first n natural numbers is an excellent one to use to illustrate a form of reasoning known as mathematical induction. Let's see how we would develop and prove Gauss' formula for adding up numbers.

To show that we are really proving a lot of separate facts, we start by listing a few of those facts, designating them as theorems. Of course, you can simply verify each of the following theorems by just doing the arithmetic. That's fine for now.

A.1 Theorem. $1 = \frac{(1)(2)}{2}$.

A.2 Theorem. $1 + 2 = \frac{(2)(3)}{2}$.

A.3 Theorem. $1 + 2 + 3 = \frac{(3)(4)}{2}$.

A.4 Theorem. $1 + 2 + 3 + 4 = \frac{(4)(5)}{2}$.

A.5 Theorem. $1 + 2 + 3 + 4 + 5 = \frac{(5)(6)}{2}$.

Okay, this is getting a little tedious. Let's see that it is not necessary to start each of this potentially infinite list of theorems from scratch. Once we have successfully proved one of these theorems, verifying the next one is much easier.

A.6 Question. *Can you use the fact that* $1 + 2 + 3 + 4 + 5 = \frac{(5)(6)}{2}$ *to verify that*

$$1 + 2 + 3 + 4 + 5 + 6 = \frac{(6)(7)}{2},$$

without having to re-add $1 + 2 + 3 + 4 + 5$?

Hopefully, your strategy did not depend in any meaningful way on the specific numbers involved. To clarify this fact, let's do another one. Notice that you are not asked to verify the sum up to 129—just accept that one as true.

A.7 Question. *Suppose it is true that* $1 + 2 + 3 + \cdots + 129 = \frac{(129)(130)}{2}$. *Can you use this fact to show that*

$$1 + 2 + 3 + \cdots + 129 + 130 = \frac{(130)(131)}{2}?$$

Try to do it without performing extensive addition.

Just one more to drive the point home.

A.8 Question. *Suppose it is true that* $1 + 2 + 3 + \cdots + 172391 =$ $\frac{(172391)(172392)}{2}$. *Can you use this fact to show that*

$$1 + 2 + 3 + \cdots + 172391 + 172392 = \frac{(172392)(172393)}{2}?$$

In fact, what you are really doing is proving that if you know that the formula holds for any natural number, then it also holds for the next natural number.

A.9 Exercise. *Suppose some natural number* k *is chosen and you are told it is true that* $1 + 2 + 3 + \cdots + k = \frac{(k)(k+1)}{2}$. *Use this fact to show that*

$$1 + 2 + 3 + \cdots + k + (k + 1) = \frac{(k + 1)(k + 2)}{2}.$$

Once you have done the above exercise, you have all the ingredients to prove that the formula is true for any number. You have proved (1) that the formula is true for the first natural number and (2) that you can always take one more step, that is, you have proved that if the formula is true for any given natural number, then it is also true for the next natural number. Why do those two steps convince you that the formula must be true for all natural numbers? This reasoning provides a proof of the following theorem.

A.10 Theorem. *Let* n *be a natural number. Then* $1 + 2 + 3 + \cdots + n = \frac{(n)(n+1)}{2}$.

The strategy of (1) proving a base case and then (2) proving that the truth of the assertion of an arbitrary natural number implies its truth for the next natural number is a method of reasoning called *proof by induction*.

Another formula

Let's go through the same process for another formula. Start by directly verifying the first few theorems.

A.11 Theorem. $1 + 2 = 2^2 - 1$.

A.12 Theorem. $1 + 2 + 2^2 = 2^3 - 1$.

A.13 Theorem. $1 + 2 + 2^2 + 2^3 = 2^4 - 1$.

A.14 Theorem. $1 + 2 + 2^2 + 2^3 + 2^4 = 2^5 - 1$.

Can you use the truth of one step to prove the truth of the next one?

A.15 Question. *Can you use the fact that* $1 + 2 + 2^2 + 2^3 + 2^4 = 2^5 - 1$
to verify that
$$1 + 2 + 2^2 + 2^3 + 2^4 + 2^5 = 2^6 - 1,$$
without performing extensive arithmetic?

In the next question, don't independently verify the case up to 2^{37}—just assume that formula is true to do the next higher case.

A.16 Question. *Suppose it is true that* $1 + 2 + 2^2 + \cdots + 2^{37} = 2^{38} - 1$.
Can you use this fact to show
$$1 + 2 + 2^2 + \cdots + 2^{38} = 2^{39} - 1?$$

Do it without performing any extensive arithmetic.

Of course, your method did not depend on the particular number 37, so let's write down the fact that you can now prove that you can always take one more step, that is, the truth of the formula for one natural number implies the truth of the formula for the next natural number.

A.17 Question. *Suppose it is true that* $1 + 2 + 2^2 + \cdots + 2^k = 2^{k+1} - 1$.
Can you use this fact to show
$$1 + 2 + 2^2 + \cdots + 2^k + 2^{k+1} = 2^{k+2} - 1?$$

Again, you have proved (1) that the formula is true for the first natural number and (2) that you can always take one more step, that is, you have proved that if the formula is true for any given natural number, then it is also true for the next natural number. Why do those two steps convince you that the formula must be true for all natural numbers? This reasoning provides a proof of the following theorem.

A.18 Theorem. *For every natural number n,* $1 + 2 + 2^2 + \cdots + 2^n = 2^{n+1} - 1$.

On your own

Prove the following theorems by induction.

A.19 Theorem. *For every natural number n,*
$$1^2 + 2^2 + \cdots + n^2 = \frac{n(n+1)(2n+1)}{6}.$$

A.20 Theorem. *For every natural number $n > 3$, $2^n < n!$.*

A.21 Theorem. *For every natural number n,*

$$1^3 + 2^3 + \cdots + n^3 = (1 + 2 + \cdots + n)^2.$$

Strong induction

In this section we are going to introduce a slightly different mode of reasoning that is called *strong induction*.

Consider the following game involving two players, whom we will call Player 1 and Player 2. Two piles each containing the same number of rocks sit between the players. At each turn a player may remove any number of rocks (other than zero) from *one* of the piles. The player to remove the last rock wins. Player 1 always goes first.

A.22 Theorem. *If each pile contains exactly one rock, Player 2 will win.*

A.23 Theorem. *If each pile contains two rocks, Player 2 has a winning strategy.*

A.24 Theorem. *If each pile contains three rocks, Player 2 has a winning strategy.*

A.25 Theorem. *If each pile contains four rocks, Player 2 has a winning strategy.*

A.26 Question. *In proving the theorem for piles with four rocks each, did you consider all possible scenarios, or did you make use of the previous three theorems?*

In the next question you are not being asked to analyze each of the first 11 cases. Instead, you are asked to assume that those have been done and then use that information to show that Player 2 has a winning strategy when there are 12 rocks.

A.27 Exercise. *Suppose you know that Player 2 has a winning strategy for this game when the number of rocks in each pile is 1, 2, 3, ..., 10, or 11. Show that Player 2 has a winning strategy when each pile contains 12 rocks.*

Of course, the number 11 could have been any number. Let's replace it with a variable.

A.28 Exercise. *Let k be a natural number. Suppose you know that Player 2 has a winning strategy for this game when the number of rocks in each pile is any one of the following natural numbers:* 1, 2, 3, ..., k. *Show that Player 2 has a winning strategy when each pile contains* k + 1 *rocks.*

You have proved (1) that Player 2 has a winning strategy for the first natural number and (2) that you can always take one more step, that is, you have proved that if Player 2 has a winning strategy for each natural number up to a certain point, then Player 2 has a winning strategy for the next natural number. Why do those two steps convince you that Player 2 has a winning strategy for any size of beginning piles? This reasoning provides a proof of the following theorem.

A.29 Theorem. *For any natural number n of rocks in each pile to begin, Player 2 has a winning strategy.*

The strategy of (1) proving a base case and then (2) proving that the truth of the assertion for all natural numbers up to a certain natural number implies its truth for the next natural number is a method of reasoning called *proof by strong induction.*

On your own

Prove the following theorems by strong induction.

A.30 Theorem. *Every natural number can be written as the sum of distinct powers of* 2.

A.31 Theorem. *Every natural number greater than* 7 *can be written as a sum of* 3*'s and* 5*'s.*

Definition. A polynomial is said to be *reducible* if it can be written as a product of two polynomials each of smaller degree. Otherwise it is said to be *irreducible.*

A.32 Theorem. *Every polynomial can be written as a product of irreducible polynomials.*

A.33 Exercise. *Describe in detail the strategies of induction and strong induction and explain why those modes of proof are valid.*

Index

About the Authors

David Marshall was born in Anaheim, California and spent most of his early life in and around Orange County. After receiving a bachelor's degree in mathematics from California State University at Fullerton he left the Golden State to attend graduate school at the University of Arizona. David received his Ph.D. in mathematics in 2000, specializing in the field of algebraic number theory. He held postdoctoral positions at McMaster University in Hamilton, Ontario and The University of Texas at Austin before becoming an Assistant Professor at Monmouth University in West Long Branch, New Jersey. David has been an active member of the MAA and AMS for over 10 years and currently serves as the Program Editor for the MAA's New Jersey Section.

Edward Odell was born in White Plains, New York. He attended the State University of New York at Binghamton as an undergraduate and received his Ph.D. from MIT in 1975. After teaching 2 years at Yale University he joined the faculty of The University of Texas at Austin where he has been since 1977, currently as the John T. Stuart III Centennial Professor of mathematics. He is an internationally recognized researcher in his area, the geometry of Banach spaces and is a much sought after speaker. Odell was an invited speaker at the 1994 International Congress of Mathematicians in Zurich. He has given series of lectures at various venues in Spain and recently at the Chern Institute in Tianjin, China. He is the co-author of *Analysis and Logic* and the co-editor of two books in the Springer Lecture series.

Michael Starbird received his B.A. degree from Pomona College and his Ph.D. from the University of Wisconsin, Madison. He is a Distinguished

Teaching Professor of mathematics at The University of Texas at Austin and is a member of UT's Academy of Distinguished Teachers. He has won many teaching awards, including the 2007 Mathematical Association of America Deborah and Franklin Tepper Haimo National Award for Distinguished College or University Teaching of Mathematics. He has written two books with co-author Edward B. Burger: *The Heart of Mathematics: An invitation to effective thinking* and *Coincidences, Chaos, and All That Math Jazz: Making Light of Weighty Ideas.* Starbird has produced four video courses for The Teaching Company: *Change and Motion: Calculus Made Clear*; *Meaning from Data: Statistics Made Clear*; *What are the Chances? Probability Made Clear*; and, with collaborator Edward B. Burger, *The Joy of Thinking: The Beauty and Power of Classical Mathematical Ideas.*